Magnesium Alloys Structure and Properties

Edited by Tomasz Tański and Paweł Jarka

Published in London, United Kingdom

IntechOpen

Supporting open minds since 2005

Magnesium Alloys Structure and Properties
http://dx.doi.org/10.5772/intechopen.87737
Edited by Tomasz Tański and Paweł Jarka

Contributors
Ranjan Kumar Bhuyan, Bhagban Kisan, Santosh Kumar Parida, Soumya Patra, Sunil Kumar, Manjaiah Mallaiah, Gajanan M. Naik, Ravindra Ishwar Badiger, Sachin Bandadka, Narendranath Sannayellappa, Nagaraj Chelliah Machavallavan, Rishi Raj, M.K. Surappa, Katarzyna Cesarz-Andraczke, Rafał Babilas, Aneta Kania, Katarzyna Anna Młynarek, Aqeel Abbas, Song-Jeng Huang, Veeramanikandan Rajagopal, Nouha Loukil, Romana Ewa Śliwa, Moara Marques De Marques de Castro, Débora Ribeiro Lopes, Leonardo Viana Dias, Prince Sharma, Nushrat Naushin, Sahil Rohila, Abhishek Tiwari

First published in London, United Kingdom, 2022 by IntechOpen
IntechOpen is the global imprint of INTECHOPEN LIMITED, registered in England and Wales, registration number: 11086078, 5 Princes Gate Court, London, SW7 2QJ, United Kingdom
Printed in Croatia

British Library Cataloguing-in-Publication Data
A catalogue record for this book is available from the British Library

Additional hard and PDF copies can be obtained from orders@intechopen.com

Magnesium Alloys Structure and Properties
Edited by Tomasz Tański and Paweł Jarka
p. cm.
Print ISBN 978-1-83962-458-2
Online ISBN 978-1-83962-467-4
eBook (PDF) ISBN 978-1-83962-468-1

For EU product safety concerns:
IN TECH d.o.o., Prolaz Marije Krucifikse Kozulić 3, 51000 Rijeka, Croatia, info@intechopen.com or visit our website at intechopen.com.

Meet the editors

Prof. Tomasz Tański is the Head of the Department of Engineering Materials and Biomaterials, Silesian University of Technology, Gliwice, Poland, and a member of the Polish Academy of Sciences, Committee of Metallurgy. He is a specialist in non-ferrous alloys, composite materials, and nanostructured, manufacturing engineering. He has authored or co-authored more than 400 scientific publications worldwide, including 15 monographs and books and more than 116 publications on the Thomson Scientific Master Journal List. He has won twenty national and international awards. He is and has been a supervisor or contractor for more than fifteen research and didactic projects in Poland and abroad. He is a reviewer and promoter of numerous scientific papers, including eight doctoral theses in the field of nanotechnology and materials.

Paweł Jarka obtained his Ph.D. from the Department of Engineering Materials and Biomaterials, Silesian University of Technology, Gliwice, Poland, where he is currently employed. The subject of his doctoral thesis was "The optoelectronic properties of active layers obtained by PVD methods in organic photovoltaic devices." His scientific interests concern functional materials. He has participated in many scientific international conferences. He is the co-author of several dozen scientific papers and five chapters in monographs. He is also a co-editor of *Electrospinning Method Used to Create Functional Nanocomposites* and a co-author of one patent.

Contents

Preface

This book includes scientific descriptions of the latest knowledge in the field of research on the structure and properties, methods of processing, and application of magnesium alloys. Magnesium alloys have a very good density-to-strength ratio, high vibration damping capacity, and high dimensional stability, making them useful for a variety of applications in the automotive, aviation, sports, electronic, medical, space, and defense industries. Additionally, these alloys have excellent castability and low casting shrinkage. Thus, magnesium alloys have captured great interest from scientists and researchers, as evidenced by the growing number of scientific publications discussing them (**Figure 1**).

Written by international experts, this book presents information on many aspects of engineering of aluminum alloys, including general information on casting technologies, the influence of alloying elements and casting conditions on structure and properties, and corrosion behavior and stress corrosion cracking mechanisms, with particular emphasis on biomedical applications. It also discusses innovative techniques for improving alloy properties using thermo-mechanical processing via the equal channel angular extrusion (ECAE) method or solidification processing via the precursor approach. Also noteworthy are the chapters describing the use of magnesium in the structures of metal matrix composites and nano-composites to develop engineering materials with a wide range of applications.

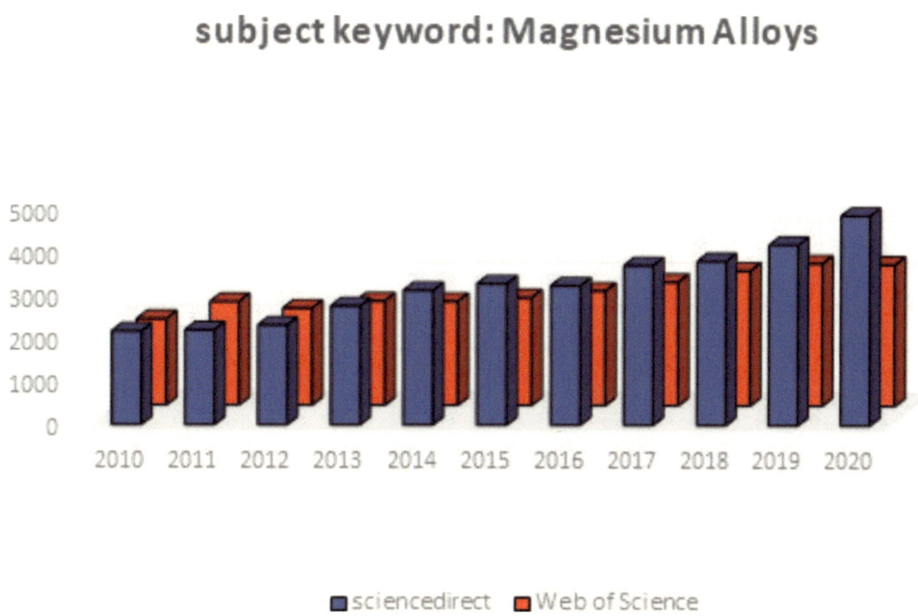

Figure 1.
Summary of the number of scientific articles with the keywords "magnesium alloys" according to Science Direct and Web of Science bases.

This book contains a compendium of knowledge on the structure and properties of aluminum alloys, taking into account the latest trends in improving their properties. It also extensively describes the use of the alloys themselves and composites based on their matrixes. It is a valuable resource for technologists from various industries and scientists in the field of engineering magnesium alloys.

Tomasz Tański
Head of the Department,
Department of Engineering Materials and Biomaterials,
Silesian University of Technology,
Gliwice, Poland

Paweł Jarka
Department of Engineering Materials and Biomaterials,
Silesian University of Technology,
Gliwice, Poland

Section 1

Forming and Processing of Magnesium Alloys with Special Properties

Chapter 1

Effect of ECAE Die Angle on Microstructure Mechanical Properties and Corrosion Behavior of AZ80/91 Magnesium Alloys

Gajanan Manjunath Naik, Sachin Bandadka, Manjaiah Mallaiah, Ravindra Ishwar Badiger and Narendranath Sannayellappa

Abstract

Magnesium alloys have poor tensile strength, ductility and corrosion resistance properties associated with other engineering materials like aluminum alloys, steels and superalloys etc. Therefore, many researchers worked on equal channel angular pressing of magnesium alloys to improve the mechanical properties and corrosion resistance. In this work, the effect of channel angles on material properties was investigated during equal channel angular pressing of AZ80/91 magnesium alloy using processing route-R at 598 K processing temperature. Channel angles of 90^0 and 110^0, common corner angle of 30^0 have been considered for the study. It has been revealed that the channel angle has a significant influence on deformation homogeneity, microhardness, ultimate tensile strength, ductility, and corrosion behavior of AZ80/91 magnesium alloys. Specifically, AZ80/91 Mg alloys processed through 90^0 channel angle i.e. die A is considered as optimal die parameter to improve above-said material properties. Investigation showing concerning as-received AZ80 and AZ91 Mg alloy indicates 11%, 14% improvement of UTS and 69%, 59% enhancement in ductility after processing through 4P through die A (90°) at 598 K respectively. Also, the corrosion rate reduces to 97% and 99% after processing the sample with 4P-ECAP die A (90°) at the same processing temperature for AZ80 and AZ91 Mg alloys respectively. This is mainly due to grain refinement and distribution of $Mg_{17}Al_{12}$ secondary phase during ECAP.

Keywords: ECAP, magnesium alloys, microstructure, mechanical properties, corrosion resistance

1. Introduction

Wrought magnesium alloys are one of the most promising lightweight materials of special interests in structural applications due to their homogeneous microstructure and improved mechanical properties compared to as-cast Mg alloys [1]. Mg alloys are one of the most reactive metals that have poor corrosion resistance and low mechanical properties, which limit its applications in industries. Therefore enhancement of mechanical properties and corrosion resistance has led to greater interest in magnesium alloys because of its special applications [2–4]. Presently much effort is required for preparation of magnesium alloys with a grain size lower than 1 μm, i.e. ultrafine-grained (UFG) materials to improve the strength and corrosion resistance of Mg alloys, many researchers worked and finally developed a severe plastic deformation (SPD) process which greatly contributes towards grain refinement and distribution of secondary phases to enhance mechanical and corrosion properties [5]. However, in SPD, ECAP is most developed and frequently used metal-working technique for significant materials hardening due to increasing dislocation density and considerable grain size reduction to sub-micro-level [6]. Finally, ultra-fine grain structure and uniformly distributed secondary phase particles increase re-passivation tendency, which exhibits the improved mechanical properties and corrosion resistance. The ECAP process was planned with two equal channels: traversing at particular angles called the die channel angle (ϕ) and the corner angle (ψ) subtended at the channels' intersection. In this work, the effect of ECAP die channel angles and processing temperature on microstructures, mechanical properties and corrosion resistance of ECAPed AZ80/91 Mg alloys were investigated.

2. Experimental work

2.1 Wrought AZ80/91 series magnesium alloys

Wrought AZ80 and AZ91 commercially available magnesium-based alloy, was selected as a workpiece material because of its high strength and low cost when compared to other magnesium-based alloys. The Mg alloy was procured in the form of a rod with dimensions 18 mm diameter and 200 mm length, from Exclusive Magnesium Pvt. Limited, Hyderabad, India. The chemical composition of AZ80/91 Mg alloys. Also, chemical composition presented along with the micro-structure shown in **Figures 1** and **2**.

2.2 Equal channel angular extrusion

Die-Material: Hot Die steel (HDS) was used for die making. The ECAE die design was done by using solid edge V.7 software and fabrication was carried out at Government Tool Room and Training Centre (GTTC) Baikampady, Mangalore, Karnataka, India. **Figure 3** depicts the ECAE die having 90^0 and 110^0 channel angle

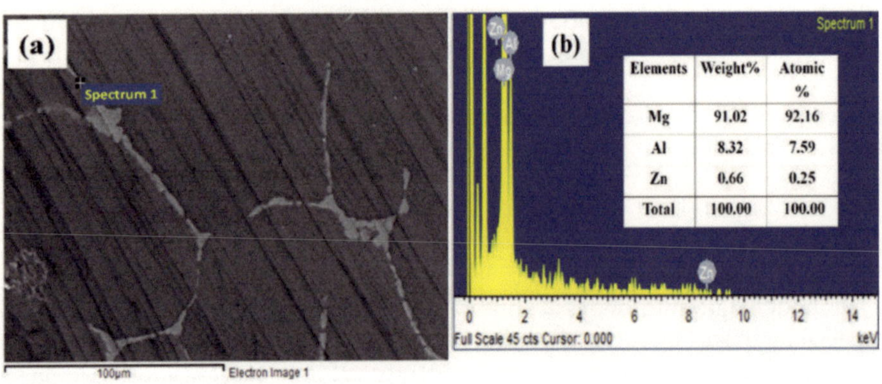

Figure 1.
AZ80 Mg alloy (a) microstructure (b) EDS results.

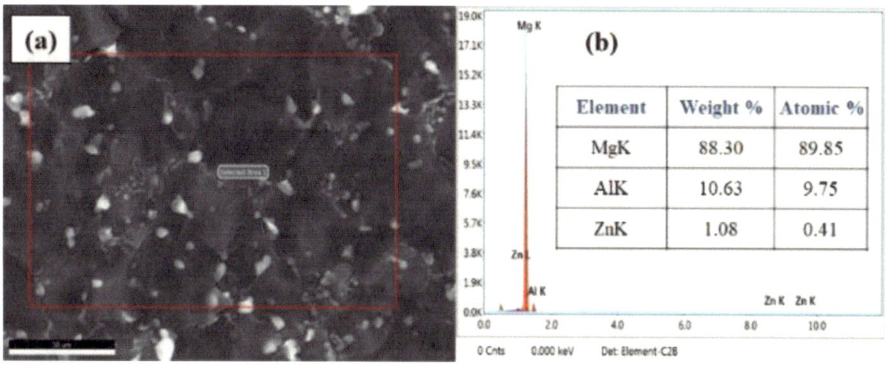

Figure 2.
AZ91 Mg alloy (a) microstructure (b) EDS results.

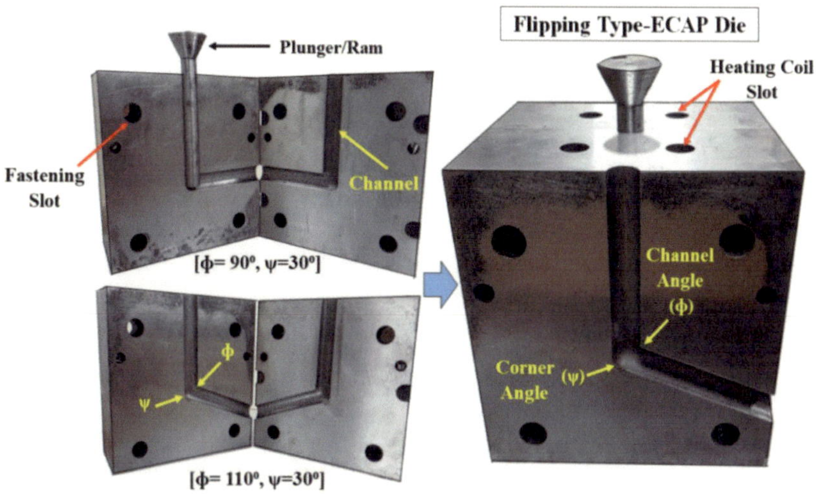

Figure 3.
ECAP die having 90° and 110° channel angle and 30° corner angle (Naik et al. Copyrights: Diary number: 14668/2018-CO/L., Reg. No: L-79923/2018).

and 30^0 corner angle. An equal channel angular extrusion thermo-mechanical apparatus is provided grain refinement to improve mechanical properties and corrosion resistance of magnesium alloys. ECAE is promising extrusion technique to achieve ultra-fine equiaxed grains without changing the shape of the workpiece. The extrusion die assembly having two angles such as channel angle (ϕ) and corner angle (ψ) which decides the induced shear strain during ECAP. Current study focusses on the effect of ECAE die angles on microstructure, mechanical properties and corrosion resistance of AZ80/91 Mg alloys. In the present study route, *R* was considered to deform the material to achieve a fine grain structure. Route *R*: the specimen is inverted to the original position in each ECAP passes. The experimental setup of equal channel angular extrusion is shown in **Figure 4**. During operation ECAP die (**Figure 4**) is placed on the base plate of UTM and plunger is aligned exactly to the center of die where the specimen is placed inside the channel for pressing. The sample is placed in the die channel and needs to make sure that the temperature of the specimen also reached the 598 K.

A lubricant as Molybdenum disulfide (MoS_2) was used to minimize the frictional effects between samples and die. After attaining the required

Figure 4.
Equal channel angular extrusion experimental setup.

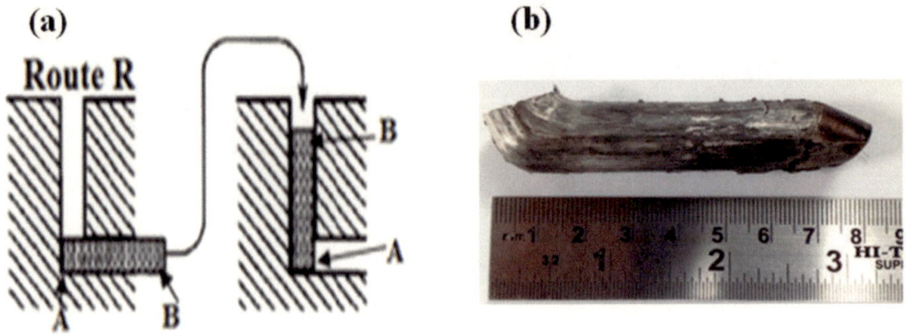

Figure 5.
(a) Route-R and (b) ECAPed sample.

temperature, the sample is pressed by applying a load at the rate of 1 mm/sec ram speed using plunger attached to the UTM for deformation of the specimen. The channels are intersected to impose the total strain on the material to get a fine grain structure. This process is repeated by using route R, where the samples were inverted from its initial position between two successive passes as shown in **Figure 5(a)**. The processed samples after ECAP operation is shown in **Figure 5(b)** [7–12].

2.3 Microstructural characterization

This section focuses on all characterization techniques of ECAPed magnesium alloys including all forms of microscopy and analysis especially microanalysis and surface analytical techniques. The methodology of microstructure analysis and phase identification are also discussed.

2.3.1 Optical microscope

Specimens for microstructure inspection were prepared by mechanical polishing with silicon carbide abrasive papers (grades of 400, 800, 1000, 1200, 1500, 2000) followed by cloth polishing using diamond paste and kerosene for obtaining mirror finish surface and finally cleaned with acetone. Further, Etching was carried on the polished surface for approximately 3 to 5 s, in a solution of 4.2 g picric acid, 10 ml acetic acid, 10 ml distilled water and 70 ml ethanol for 3–5 s [8–12]. So that sample turns light brown and washed with running distilled water and dried. Microstructures and elements distributions were observed and analyzed using optical microscopy by image analyzer facilitated BIOVIS material plus software, and average grain sizes were measured by linear intercepts method according to ASTM E-112. Same samples were observed under a scanning electron microscope.

2.3.2 Scanning electron microscopy

Scanning electron microscope (SEM) equipped with energy dispersive spectroscopy (EDS) Backscattered electron (BSE) detector coupled with the EDS allows for composition identification of materials. Scanning Electron Microscopy Model: JEOL JSM–638OLA from JEOL, USA, operated at 30 kV; Magnification range-3,00,000×, which allows studying the microstructures and surface morphologies.

2.3.3 X-Ray diffraction

X-ray diffraction (XRD) is one of the primary techniques used for the characterization of crystalline solids and determination of their structure or phases. XRD measurements are carried out in M/s Proto Manufacturing Ltd., CANADA make PROTO–iXRD MGR40, wherein the analysis was carried out 2θ: an angular range of 20° to 90° at a scanning speed of 2^0/min. The XRD patterns obtained were analyzed with the help of PCPDFWIN software to identify the formation of primary, secondary and ternary phases.

2.4 Mechanical testing

2.4.1 Microhardness test

Microhardness test is also performed on as-received heat-treated and some of the deformed or ECAPed samples. The measurements were carried out at a load of 100gm and dwell of 13 s the microhardness was calculated using the expression [13]. Microhardness Model: MVH–S–AUTO from OMNI TECH, PUNE, INDIA.

2.4.2 Tensile test

The tensile test is used to evaluate the strength and ductility of as-received and equal channel angular extruded sample. Specimens were prepared according to the ASTM-E8 standard with 16 mm gauge length. The tensile properties of magnesium alloys were measured using UTM-Shimadzu AG-X plus™ equipped with 100 kN load cell and operated with a steady cross-head speed of 0.25 mm/min during all the tensile tests. Three samples were tested for each condition and uniaxial tensile testing was accomplished at room temperature and average reading was calculated and presented.

2.5 Electrochemical corrosion test

Corrosion study of AZ80/91 wrought Mg alloys was investigated using electrochemical corrosion analyzer, model: Gill AC-1684, supplied by Tech-science Pvt. Limited, Pune (India). The potentiodynamic polarization tests were conducted in 3.5 wt.% NaCl solution to estimate corrosion resistance or rate of corrosion of AZ80/91 wrought alloys. The auxiliary electrode (AE) was made of graphite (Gr) and the reference electrode (RE) was made of a saturated calomel electrode (SCE). $1cm^2$ area of the working electrode (AZ80/91 alloys) was exposed to the 3.5 wt.% NaCl solution. Before the electrochemical corrosion test, specimens were polished with 600, 800, 1000, 1200, 1500, 2000 grit emery papers and washed with ethanol. The specimens were kept in corrosion cell kit in NaCl solution for 20 min to stabilize the open circuit potential (OCP). Further, the AC impedance test of starting frequency 10 kHz and ending frequency 10 MHz with a scan speed of 5 mV/s and cyclic sweep experiments with −250 to +250 mV was carried on the electrochemical analyzer. Surface morphology of the corroded samples was examined by SEM. The corrosion product was removed using 200 g/L of chromic acid and 10 g/L of $AgNO_3$ solutions. The corrosion rate of the alloy was calculated by using Eq. (1).

$$\mathrm{CR}\left(\mathrm{mm}/\mathrm{y}\right)=3.27X\,10^{-3}\,\frac{i_{corr}\;X\,A}{\rho} \tag{1}$$

where CR is the corrosion rate in miles per year, A is the molar mass (for magnesium 24.3 g/mol), I_{corr} is the corrosion current density in $\mu A/cm^2$, n is the valance and ρ is the density (1.74 g/cm^3).

3. Results and discussion

3.1 Effect of ECAP die channel angle on AZ80/91 magnesium alloy

So far, many simulation studies have been executed to examine the impact of different die parameters on deformation homogeneity, strain rate, workflow etc. Although many researchers have been carried out on the efficiency of ECAP process routes and influences of various ECAP parameters on the strain behavior [14], there is limited work on a study of the effect of channel angle on grain size and other material properties through experimentally. In this chapter, the effect of ECAP channel angle on grain size, microhardness, tensile behavior and corrosion rate for different passes were analyzed using working temperatures of 598 K Furthermore, die A was used for examining above said material properties since this die gives the best results.

3.1.1 Microstructure evolution of AZ80 Mg alloy

The optical microstructures of as-received, homogenized at 673 K-24 h sample and those after ECAP processed specimens are shown in **Figures 6** and **7**. The microstructure of the as-received AZ80 Mg alloy presents the α-Mg and β-$Mg_{17}Al_{12}$ secondary phases along the grain boundaries indicated in **Figure 6(a)**. After homogenized at 673 K for 24 h secondary phases were partially dissolved along the grain boundaries as shown in **Figure 6(b)** this partial dissolution of secondary phases was achieved before ECAP and this sample is designated as 0P specimen. **Figure 7** presents the optical images of the ECAPed AZ80 Mg alloy processed

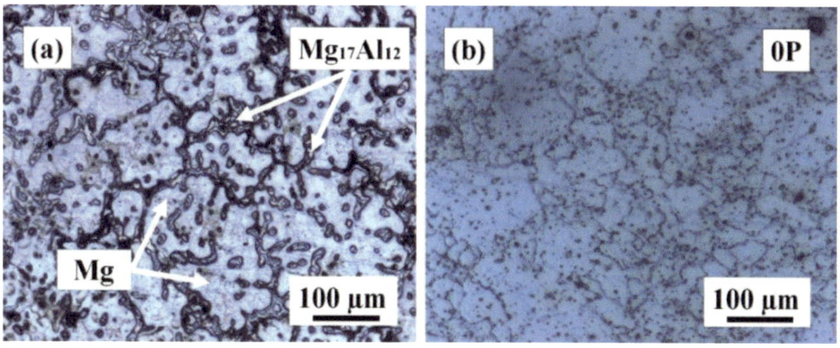

Figure 6.
Optical images of (a) as-received (b) homogenized at 673 K-24 h.

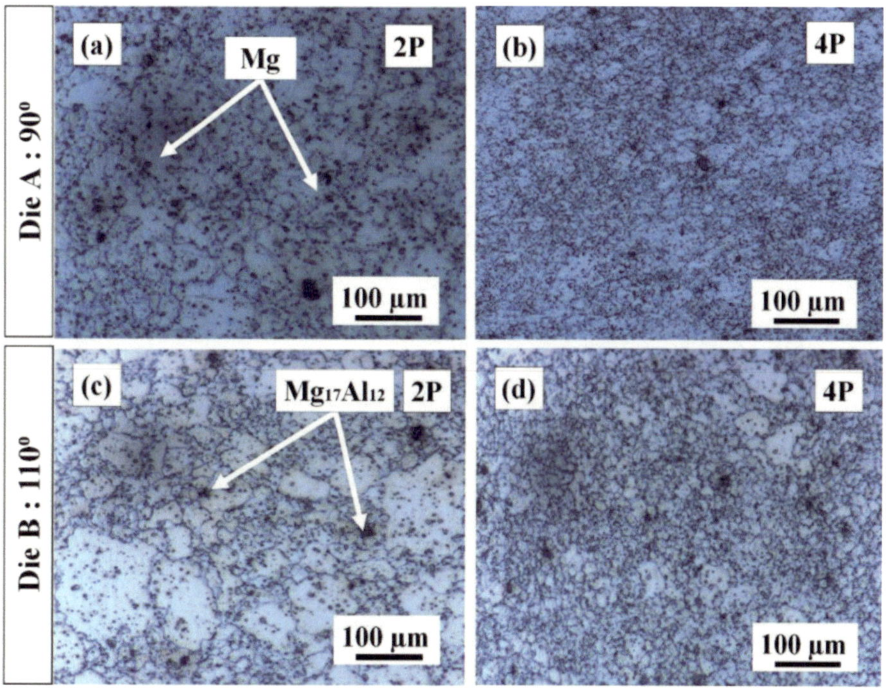

Figure 7.
Optical images for die A: (a) 2P (b) 4P and die B: (c) 2P and (d) 4P ECAP passes.

through two ECAP die of 2 and 4 passes at 598 K processing temperature, in which the white and black contrast within the grains and along the grain boundaries represents α-Mg primary phase and β-$Mg_{17}Al_{12}$ secondary phases respectively. Also, the presence of α-Mg and β-$Mg_{17}Al_{12}$ phase in AZ80 alloys was confirmed through by the XRD analysis shown in **Figure 8**. The microstructure of the ECAPed Mg alloy showed significant grain refinement and bi-modal grains after ECAP of two passes for both die A and B, as shown in **Figure 7(a)** and **(c)**. These heterogeneous grains were typically obtained under the condition of lower deformation. When ECAP passes were gradually increased up to four passes bi-modal grain structure disappeared due to a large amount of induced plastic strain, as a result of the average grain size of ECAP-4P through die A was ~6.35 μm and the secondary phases are uniformly distributed throughout the material as shown in **Figure 7(b)**. Whereas ECAP-4P processed through die B exhibited slightly larger grains compared to die A, the obtained grain size is of about ~9.77 μm. Hence, the effectiveness of grain refinement can be enhanced based on a channel angle, particularly, material processed through 90° channel angle exhibited better grain refinement.

3.1.2 Microstructure evolution of AZ91 Mg alloy

Figure 9 shows the microstructure of the as-received and homogenized AZ91 Mg alloy. The microstructure of the as-received AZ91 Mg alloy shows a coarse α-Mg phase and β-$Mg_{17}Al_{12}$ secondary phase along the grain boundaries which is confirmed through XRD analysis as shown in **Figure 8**.

The mean grain size of as-received Mg alloy is ~58.69 μm as shown in **Figure 9(a)**, measured by the linear intercept method (ASTM E 112). From **Figure 9(b)**, it could be found that after homogenization treatment at 673 K for 24 h the slight increase of mean grain size of Mg alloy of ~59.82 μm was observed this is due to grain growth

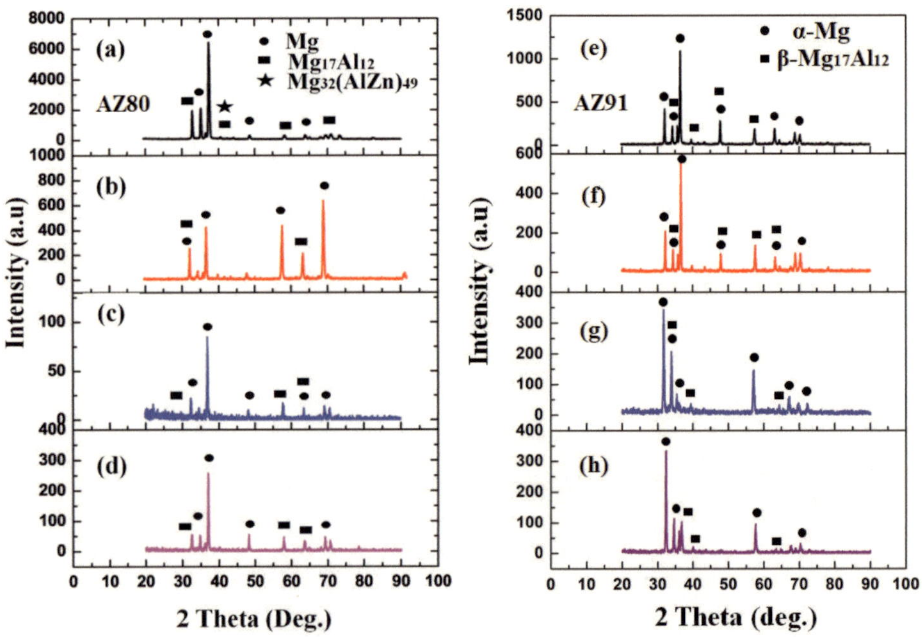

Figure 8.
XRD analysis on AZ80 Mg alloys (a) as-received (b) homogenized at 673 K-24 h (c) die A: 2P at 598 K (d) die A: 4P at 598 K. XRD analysis on AZ91 Mg alloys (e) as-received (f) homogenized at 673 K-24 h (g) die A: 2P at 598 K and (h) die A: 4P at 598 K.

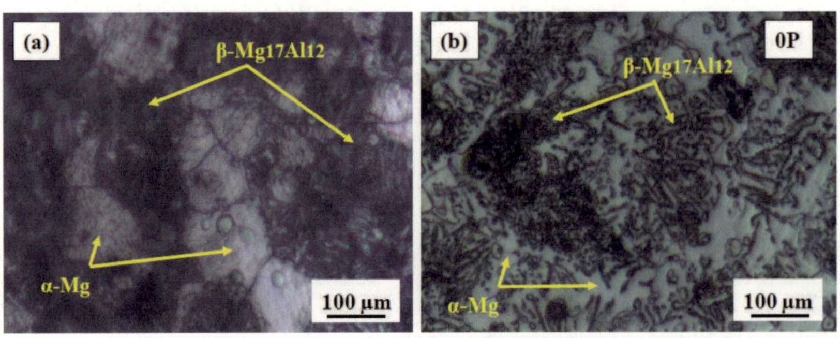

Figure 9.
Optical images of (a) as-received (b) homogenized at 673 K-24 h.

effect during the homogenization process. Similar observations are made by Nikulin et al. [15]. After ECAP, the microstructure of the alloy is effectively refined by dynamic recrystallization process (DRX) [16]. From **Figure 10**, it was observed that two-pass pressing through die A and die B exhibited bimodal grain structure and more fine grains appeared for the processing in the 90^0 die than the 110^0 die. The effectiveness of the grain refinement was observed after four passes of pressing in the 90^0 die as shown in **Figure 10(b)**. Also, it is noticed that with the increase in the number of ECAP passes, the amount of fine grains is increased greatly.

3.1.3 Variation of grain size with two different die after 2 and 4 ECAP passes for AZ80/91 Mg alloys

Figure 11 displays the variation of average grain size of processed and unprocessed AZ80 and AZ91 Mg alloys of 2 and 4 ECAP passes through die A and die B. As it could be observed from **Figures 6** and **9** as-received and homogenized (0P) Mg alloy has moderately large grain size approximately ~50.20 μm and ~50.70 μm for AZ80 and 58.69 μm, 59.82 μm for AZ91 alloy respectively. The increased average grain size of Mg alloy after homogenization treatment AZ80/91 Mg alloy at 673 K-24 h is due to the phenomenon of grain growth effect [8, 17, 18]. Further, it can be shown that after 2P and 4P ECAP volume fraction of grains increases compared to as-received and homogenized Mg alloy. Mean grain size of AZ80 Mg alloy after ECAP-2P and 4P were ~28.87 μm and ~6.35 μm respectively for die A. Similarly, the average grain size of same Mg alloy processed through die B is ~36.14 μm and ~9.77 μm for ECAP-2P and 4P respectively. Further, an average grain size of AZ91 Mg alloy after 2P and 4P of ECAP were ~30.86 μm, ~7.58 μm for die A and ~36.14 μm, ~9.7 μm for die B respectively. It is apparent that the obtained grain refinement is due to DRX during ECAP and they increase in many ECAP passes which result in much smaller grain structure. However, from this it is noticed that the alloy processed with 90^0 die shows smaller grain sizes than 110^0 die for both alloy, this is due to

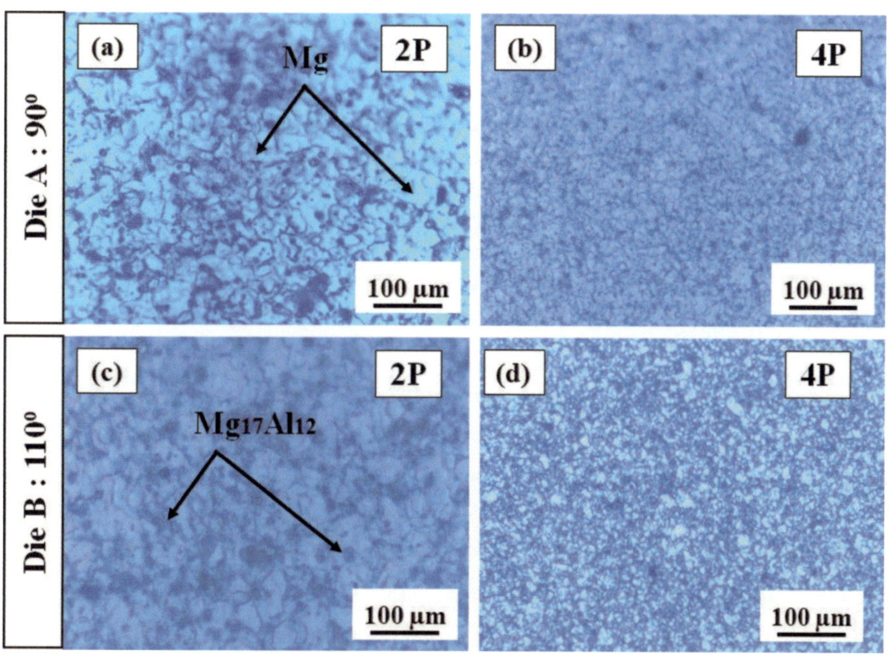

Figure 10.
Optical images for die A: (a) 2P (b) 4P and die B: (c) 2P and (d) 4P ECAP passes.

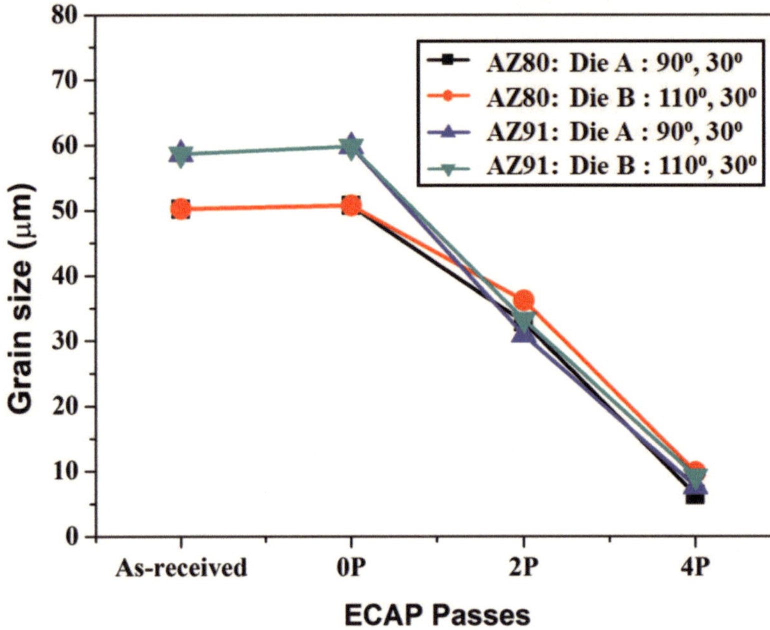

Figure 11.
Variation of average grain size with two different die.

the accumulation of very large plastic strain while processing with a low angle die [19, 20]. The calculated equivalent plastic strain for 110^0 to be ~0.742 and ~1.015 for 90^0 indeed, the strain developed by 4P ECAP through 90^0 die is higher than that of 110^0 die. Therefore, large strain in the material exhibited more dislocation density lead to the formation of fine grains during this process. Therefore, undoubtedly it is evident that ECAP die angles significantly affect the deformation homogeneity and this influences the variation in microstructure [21–23].

Also, the microstructural change contributes towards improved mechanical properties and corrosion resistance. Finally, in general, AZ Mg alloy processed through die A and die B showed the same trend of decreasing grain size from the homogenized condition. By extruding in the die A, the mean grain size of AZ80 and AZ91 Mg alloy decreased by 35% and 22% when compared with material processed through die B [19, 20]. Also, from the result, it was observed that AZ80 Mg alloy processed through die A at 598 K exhibited fine grain structure of about ~6.35 μm after four ECAP passes, which is lower when compared to ECAPed AZ91 Mg alloy processed at same processing temperature.

3.2 X-ray diffraction analysis

The X-ray diffraction patterns of AZ80/91 Mg alloy before and after ECAP processes as shown in **Figure 8**. The XRD patterns of the as-received, homogenized at 673 K and ECAPed AZ Mg alloys revealed two sets of peaks, one for the α-Mg primary phase and another one for the β-$Mg_{17}Al_{12}$ secondary phase. But as-received alloy of AZ80 has shown new peaks corresponding to the formation of the ternary phase appeared at 41.4° as shown in **Figure 8(a)** which is disappeared after homogenization treatment and ECAP depicts in **Figure 8(b)–(d)** due to diffusion annealing treatment and dynamic precipitation during the ECAP process. Further, **Figure 8(c)** and **(d)** presents the XRD patterns for ECAPed AZ80 Mg alloys for 2P and 4P processed with die A at processing temperature 598 K. It was observed that the peak intensities were increased after 4P ECAP when compared to the ECAP-2P sample. This is due to an increased volume fraction of secondary phases and more homogenous microstructure. But 2P ECAP processed sample presented lower peak intensity this is mainly due to non-homogeneity in the microstructure and crystal defects.

Furthermore, **Figure 8(e)–(h)** shows the XRD spectra of AZ91 Mg alloy (e) as-received (f) the homogenized at 673 K for 24 h (g) the two-passed AZ91 Mg alloy ECAPed with die A at 598 K and (h) the four-passed AZ91 Mg alloy ECAPed with die A at 598 K. Regardless of the number of ECAP pass, the as-received and processed samples contained α-Mg and β-$Mg_{17}Al_{12}$ phase. The intensity of the peak in the ECAP processed specimens at 598 K is lower than that of the as-received specimen. Also, it can be seen that there exists great difference on the magnitude of the peak intensity of ECAP processed specimen at 598 K for two and four passes this is mainly due to induced plastic strain during ECAP similar results has been observed by Avvari et al. [24–28].

3.3 Mechanical behavior

This section explains the effect of ECAP die channel angle on mechanical properties of as-received and ECAPed AZ80/91 Mg alloys.

3.3.1 Effect of die parameters and processing temperature on microhardness

Figure 12 shows the impact of channel angle on microhardness during ECAP of AZ80/91 Mg alloys. From the results, it was observed that AZ80/91 Mg alloy processed through lower channel angle of 90° (die A) exhibited enhanced micro-hardness when compared to material processed through die B at 598 K after 4 Passes of ECAP. The improved microhardness is mainly due to the accumulation of large plastic strain while processing at 90° channel angle and obtained more equiaxed microstructure.

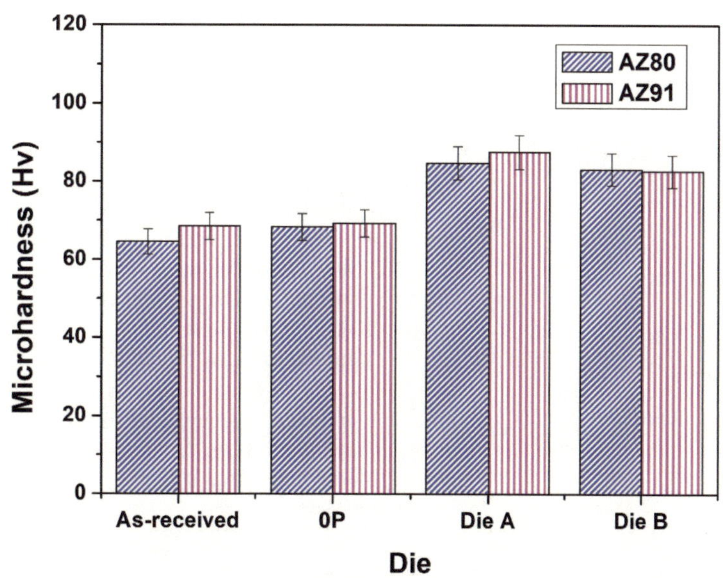

Figure 12.
Variation of microhardness for AZ80/91 Mg alloys after processing through two different dies.

Hence, die A which has 90° channel angle is considered as an optimal die parameter to get the highest Microhardness for both AZ80/91 Mg alloys. Also, from **Figure 12** it was established that there is a significant increase in Microhardness after a four pass of ECAP in AZ91 Mg alloy after processing using a die A compared to AZ80 Mg alloy processed through the same die and this is anticipated from measurements of the effective refinement of grain size.

3.3.2 Variation of tensile strength with two different die and processing temperature

The engineering stress-strain curves of un-ECAPed and ECAPed AZ80/91 Mg alloys at two different dies are shown in **Figure 13**. The stress-strain curves show that ultimate tensile strength and ductility of ECAP-4P processed AZ80 specimens at 598 K are about 489.17 MPa, and 19.03%, respectively for die A, Along with this, the same material processed through die B 4P-ECAP exhibited UTS and %elongation is 451.01 MPa and 11.76% respectively, which are higher than that of an as-received and homogenized specimen of AZ80 Mg alloys. Similarly, the AZ91 Mg alloy processed through die A has greatly improved ultimate tensile strength and ductility. Particularly, as-received AZ91 alloys have 372.74 MPa and 7.84% of UTS and ductility respectively, which is further enhanced to 432.81 MPa and 19.13% after processing through die A for 4 passes and 410.35 MPa and 13.22% of ultimate tensile strength and ductility was observed after processing through die B for 4 passes. From this result, it was found that compared with the ECAPed AZ91 Mg alloys processed with die A, the AZ80 Mg alloy processed through die A exhibited enhanced tensile properties compared to die B. This is due to an induced large amount of plastic strain during ECAP.

3.4 Corrosion study

This section illustrates the effect of ECAP die channel angle on corrosion behavior of as-received and ECAPed AZ80/91 Mg alloys. Also, presents the morphology study on corroded surfaces of as-received and ECAPed AZ80/91 Mg alloys.

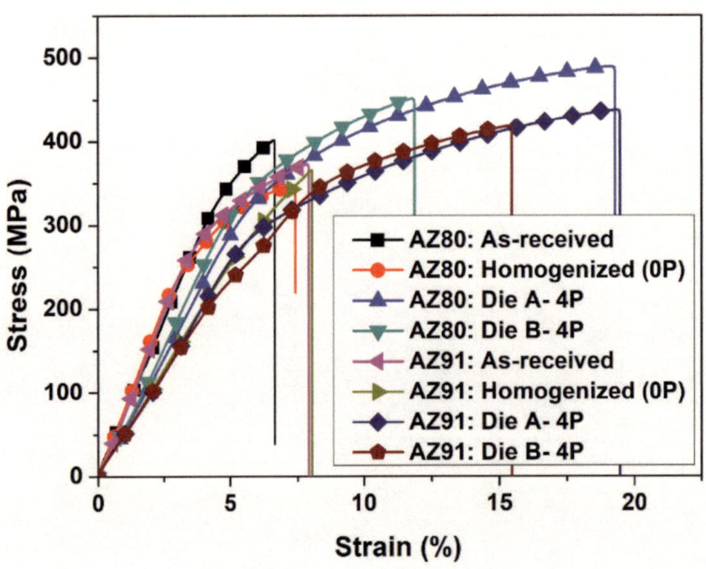

Figure 13.
Variation of tensile strength for different ECAP die.

3.4.1 Effect of ECAP die and processing temperature on corrosion rate

The corrosion results of AZ80/91 Mg alloy processed through die A and die B at 598 K after 4-ECAP passes including as-received and homogenized samples were shown in **Figure 14**.

From **Figure 14**, it was observed that the channel angle of ECAP significantly influences the grain refinement and distribution of secondary phases which contribute towards corrosion resistance. ECAP processing through die A at 598 K leads to lower corrosion rate after 4 passes of ECAP compared to die B under the same conditions for both AZ80/91 Mg alloys. This is mainly due to the lower dislocation density at recrystallization temperature [29]. Therefore the reduction of the grain size and the increase of the distribution of secondary phases can cause an improved corrosion resistance. In other words, the column chart shows the variation of corrosion rates of Mg alloys before and after ECAP process for both AZ80/91 Mg alloys. The more ECAP passes are related to the nobler corrosion potentials and the lower current density. The Mg alloy processed through die A and die B after four ECAP passes results that the ECAPed Mg processed through die A has nobler E_{corr} and I_{corr} values, leads to more corrosion resistance than the specimen extruded through die B, as-received and homogenized. Specifically, AZ91 Mg alloy processed through ECAP after 4 passes exhibited improved corrosion resistance than the ECAPed AZ80 Mg alloys this mainly due to elemental composition of AZ91 Mg alloy.

3.4.2 Effect of grain size and distribution of secondary phases on corrosion behavior

Figure 15 depicts the polarization plots for AZ80 and AZ91 Mg alloys processed with die A at 598 K. From **Figure 15(a)** it was observed that the anodic branches of the unprocessed and processed AZ80 specimen showing the continuous active dissolution of the metal this indicate that AZ80 Mg alloy exhibit poor passivity [30]. Although, E_{corr} values of ECAP processed AZ80 Mg alloys are significantly shifted to the less negative potentials and highly reduced the magnitude of I_{corr} after 2 passes.

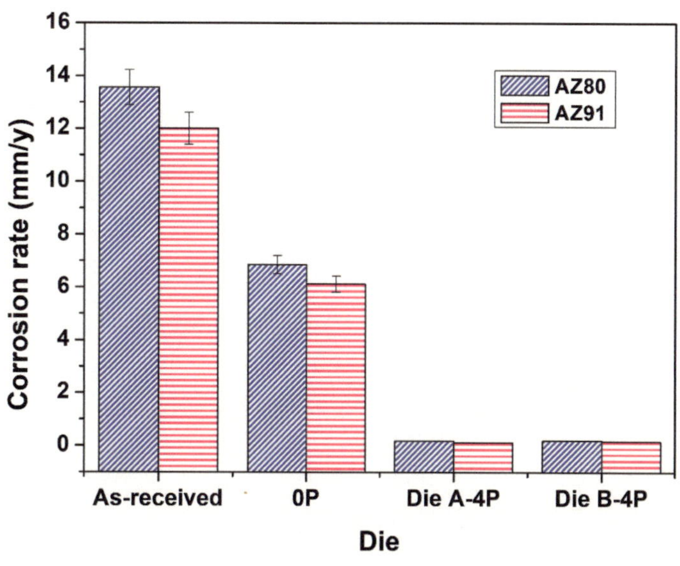

Figure 14.
Corrosion rate vs. ECAP die.

A similar observation was made by Ambat et al. [31]. Further, the polarization plot of the 4P ECAPed with 90^0 die exhibits a corrosion potential of $-1.375\ V_{SCE}$ this is higher than corrosion potential of other ECAP passes. This indicates that AZ80 Mg alloy processed with 90^0 die sample has higher pitting corrosion resistance. Moreover, polarization results specify that the ECAPed AZ80 Mg processed with 90^0 die has nobler E_{corr} values. Further, the potentiodynamic polarization curves of as-received and ECAPed AZ91 Mg specimens in 3.5 wt.% NaCl was also shown in **Figure 16(b)**. The experimental results revealed that the E_{corr}, corrosion potential of 4P-ECAPed AZ91 Mg alloys was $-1.453V_{SCE}$, which was less negative compared with the as-received alloy and other ECAP passes (**Figure 16(b)**). This phenomenon specifies that the cathodic reaction was more difficult in fine-grained Mg alloys compared to the coarse grain alloy. Therefore, with the ECAP, the corrosion potential

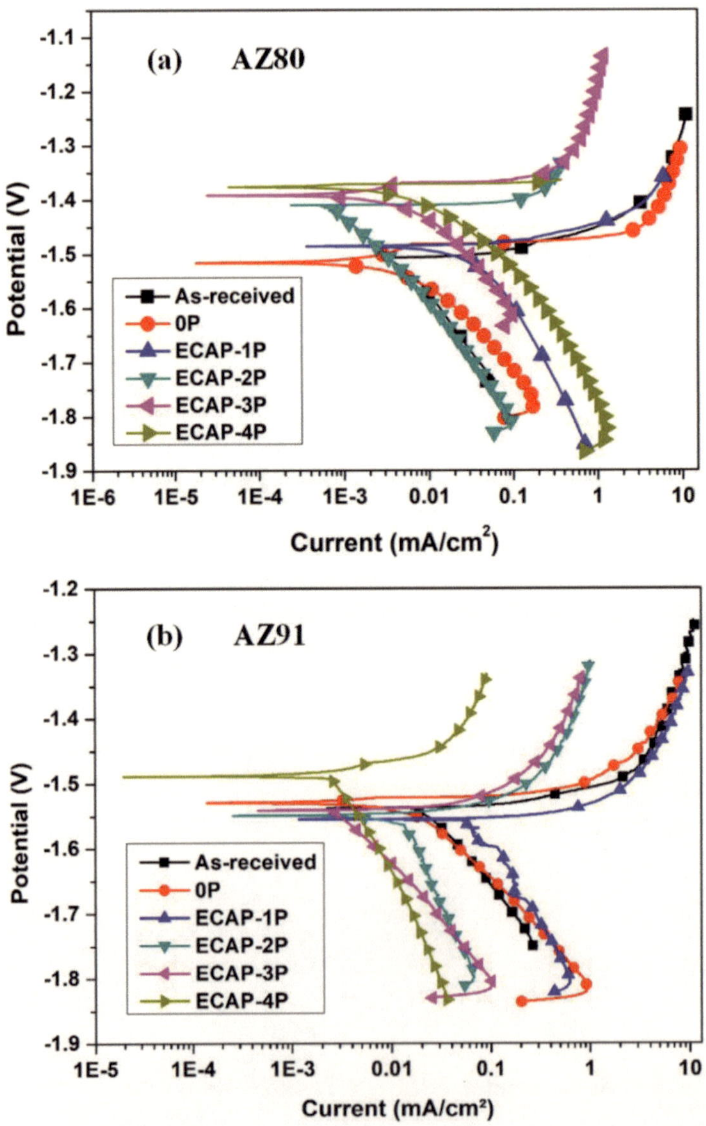

Figure 15.
Polarization curves of Mg alloys processed through die A (a) ECAPed AZ80 Mg alloys and (b) ECAPed AZ91 Mg alloys.

(E_{corr}) shifted to −1.536 V_{SCE} and −1.453 V_{SCE} after two and four ECAP passes which are considerably nobler in comparison with the as-received alloy (−1.540 V_{SCE}). However, the corrosion potential increases with the grain refinement after ECAE in the alloy. Also, the corrosion current density (I_{corr}) of 2P and 4P ECAPed AZ91 Mg alloy was 0.0173 mA/cm^2 and 0.0053 mA/cm^2 respectively, which is lesser than that of as-received AZ91 Mg alloy (0.0263 mA/cm^2). The obtained results revealed that the ECAPed Mg sample after 4 passes has nobler corrosion potential and lower current density when compared with as-received and ECAPed-2P. Therefore, ECAE increased the corrosion resistance of Mg alloy this is due to grain refinement and distribution of secondary phases, which is shown in OM and SEM microstructure in **Figure 16**. Similarly, Shahar et al. [32] explored that the grain refinement and secondary phase distribution through ECAP improves the corrosion resistance of Mg alloys.

3.5 Corrosion morphology

The corrosion morphologies of as-received and as-processed specimens of AZ80 and AZ91 Mg alloys immersed in 3.5 wt.% NaCl solution observed through scanning electron microscopy and is shown in **Figures 16** and **17**.

From **Figures 16(a)**, **(b)** and **17(a)**, **(b)**, it was observed the adequate amount of corrosion attack on the surface of the as-received and homogenized AZ80 and AZ91 Mg alloys after potentiodynamic polarization test. The ECAP performed samples of Mg alloys after the corrosion test exhibited comparatively less localized pits on the surface of ECAPed AZ80/91 Mg alloys have shown in **Figures 16(c)** and **17(c)**. This obtained result showed that pitting corrosion resistance of ECAPed Mg alloys are significantly improved through grain refinement and this is due to the distribution of secondary phases [33–35]. It is worth to declare that this appeared improved for AZ80/91 Mg alloy was due to the grain refinement, distribution of secondary phases and formation of

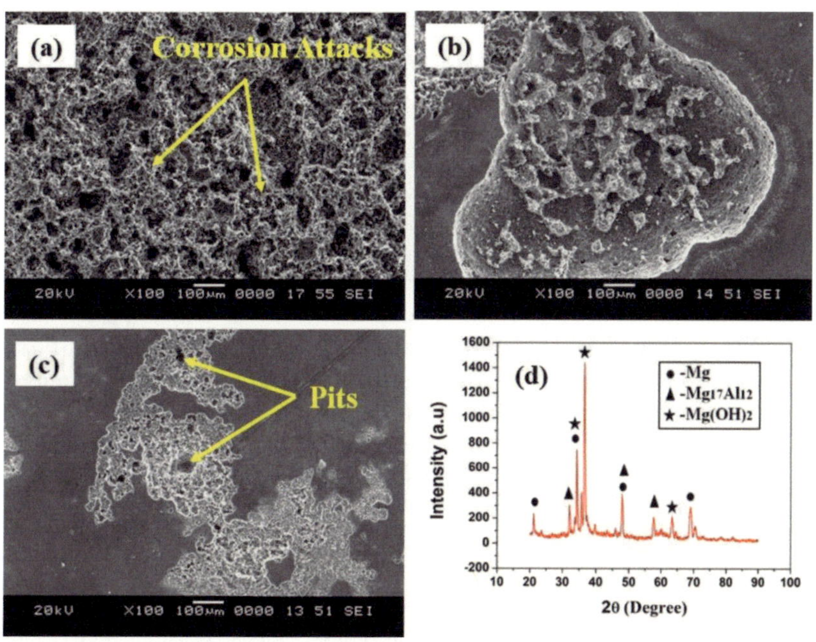

Figure 16.
Corrosion morphology of AZ80 Mg alloys (a) as-received (b) homogenized at 673 K-24 h (c) 4P-598 K and (d) XRD for a corroded specimen of as-received.

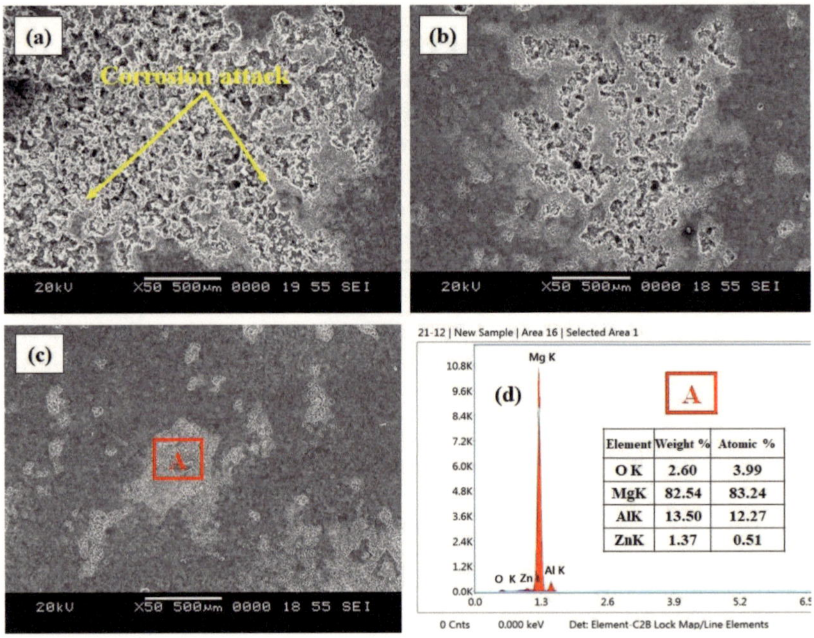

Figure 17.
Corrosion morphology of AZ91 Mg alloys (a) as-received (b) homogenized at 673 K-24 h (c) 4P-598 K and (d) EDX for corroded ECAP-4P specimen.

magnesium hydroxides formed on their surfaces which proved through microstructure and X-ray diffraction analysis shown in **Figures 16(d)** and **17(d)** for AZ80 and AZ91 Mg alloys respectively. The existence of such metal oxide partially protects the Mg surface from further dissolutions under this circumstance the breakdown of the film and the consequential nucleation and growth of a pit become more difficult. Moreover, the higher E_{corr} value for ECAPed Mg alloys at higher passes revealed that the surface of Mg alloys was more passivated against corrosion this is due to slow dissolution rate of fine grains structure [30, 31]. Finally, from the results, it was concluded that the severe corrosion attack was observed on as-received AZ80 and AZ91 Mg alloys and further continuous reduction in corrosion attack was observed for ECAPed samples. Similar kind of results and trends is reported by many authors in their studies [29, 36, 37].

4. Conclusion

AZ80/91 Mg alloy processed through both die A and die B at 598 K were discussed in this chapter. Indeed, Mg alloy processed through die A has significantly shown fine grains than die B. Since the grain size and distribution of secondary phases are a major factor in determining the strength and corrosion resistance of the material respectively, therefore die A is considered as optimal to achieve fine grain structure in our work. As a result, the fine grains obtained through die A exhibited improved mechanical and corrosion resistance discussed in the earlier sections. Also, based on the experimental results and discussion, the following conclusions were drawn.

- Increase in ECAP passes lead to homogeneous microstructure due to dynamic recrystallization which occurred during ECAP process. The secondary β-$Mg_{17}Al_{12}$ phase was reduced and uniformly distributed throughout the extruded material. Here, the effectiveness of ECAP with die A in grain refining of AZ80/91 Mg alloys was quite significant because of imposing large plastic strain of ~5.06 after four ECAP passes.

- The average grain size of AZ80 and AZ91 Mg alloy was found to be reduced to 6.35 μm and 7.58 μm respectively after processing through die A at 598 K.

- Microhardness, ultimate tensile strength and ductility for both AZ80 and AZ91 Mg alloy has been enhanced by refining grain size with an increasing number of passes. Ultimate tensile strength of the AZ91 Mg alloy decreased when compared to AZ80 Mg alloy after ECAP this is due to the presence of the secondary β-$Mg_{17}Al_{12}$ phase.

- Mechanical properties such as ultimate tensile strength, ductility and microhardness of the AZ80/91 Mg alloy are directly proportional to the ECAE passes.

- Potentiodynamic polarization test showed reduced corrosion current density (I_{corr}) which indicates higher corrosion resistance for the ECAP processed samples due to the presence of equiaxed fine grain microstructure and homogeneously distributed secondary particles ($Mg_{17}Al_{12}$).

- Polarization results showed that passive behavior of ECAPed AZ80/91 Mg sample enhances compared to as-received AZ80/91 Mg alloy owing to the grain refinement and distribution of secondary phase. An AZ80/91 Mg alloy processed with die A (90°) showed higher pitting corrosion resistance compared to die B (110°), by showing less negative pitting potential during 4P-ECAP. Also, the obtained polarization data have good agreement with the corrosion surface morphology.

Author details

Gajanan Manjunath Naik[1], Sachin Bandadka[2], Manjaiah Mallaiah[3]*, Ravindra Ishwar Badiger[4] and Narendranath Sannayellappa[5]

1 Department of Mechanical Engineering, Mangalore Institute of Technology and Engineering, Moodbidri, Mangalore, Karnataka, India

2 Department of Mechanical Engineering, Nitte Meenakshi Institute of Technology, Yelahanka, Bangalore, Karnataka, India

3 Department of Mechanical Engineering, National Institute of Technology Warangal, Telangana, India

4 Department of Mechanical Engineering, Yenepoya Institute of Technology, Moodbidri, Mangalore, Karnataka, India

5 Department of Mechanical Engineering, National Institute of Technology Karnataka, Surathkal, Mangalore, Karnataka, India

*Address all correspondence to: manjaiahgalpuji@gmail.com

References

[1] Zhongyu Cui, Xiaogang Li, Kui Xiao, Chaofang Dong. (2013). "Atmospheric corrosion of field-exposed AZ31 magnesium in a tropical marine environment," *Corr. Sci.*, 76, 243-256.

[2] Noor, E A, A.H. Al-moubaraki. (2008)."Corrosion Behavior of Mild Steel in Hydrochloric Acid Solutions." *Int. J. Electrochem. Sci.*, 3, 806-818.

[3] Hao-Miao Yang, Nan-Yi Zhang, Na Liu, Wei-Dong Xie, Xiao-Dong Peng. (2016). "Microstructure, mechanical properties, and corrosion resistance of Mg–9Li–3Al–1.6Y alloy." Rare Metals, 35, 374-379.

[4] Bao. L, Z. Zhang, Q. Le, S. Zhang, J. Cui. (2017) "Corrosion behavior and mechanism of Mg-Y-Zn-Zr alloys with various Y/Zn mole ratios," J. Alloys Comp., 712, 15-23.

[5] Shaeri M.H.,. Ebrahimi, M.T. Salehi, S.H. Seyyedein. (2016). "Materials International Effect of ECAP temperature on microstructure and mechanical properties of Al – Zn – Mg – Cu alloy." *Prog. Nat. Sci. Mater. Int.*, 26, 182-191.

[6] Brunner J.G, J. May, H.W. Höppel, M. Göken, S. Virtanen. (2010). "Localized corrosion of ultrafine-grained Al – Mg model alloys." *Electrochimica Acta*, 55, 1966-1970.

[7] Sun, H. Q., Shi, Y. N., Zhang, M. X., & Lu, K. (2007). "Plastic strain-induced grain refinement in the nanometer scale in a Mg alloy." *Acta Mater.*, 55(3), 975-982.

[8] Naik, Gajanan M., Gopal D. Gote, S. Narendranath, and SS Satheesh Kumar. (2018). "The impact of homogenization treatment on microstructure microhardness and corrosion behavior of wrought AZ80 magnesium alloys in 3.5 wt% NaCl solution." *Mater. Res. Express*, 5, 086513.

[9] Naik Gajanan, M, Narendranath, S., & Kumar, S. S. (2019). "Influence of ECAP processing routes on microstructure mechanical properties and corrosion behavior of AZ80 Mg alloy." *AIP Conf. Proc.*, 2082 (1), 030016.

[10] Naik, G. M., Gote, G. D., & Narendranath, S. (2018). "Microstructural and Hardness evolution of AZ80 alloy after ECAP and post-ECAP processes." *Mater. Today: Proc.*, 5(9), 17763-17768.

[11] Naik, G. M., Gote, G. D., Narendranath, S., & Kumar, S. S. (2018). "Effect of grain refinement on the performance of AZ80 Mg alloys during wear and corrosion." *Adv. Mater. Res.*, 7(2), 105-118.

[12] Naik, G. M., Narendranath, S., & Kumar, S. S. (2019). "Effect of ECAP Die Angles on Microstructure Mechanical Properties and Corrosion Behavior of AZ80 Mg Alloy." *J. Mater. Eng. Perform.*, 28(5), 2610-2619.

[13] Dieter, G. E., & Bacon, D. (1986). "Mechanical metallurgy" (Vol. 3). New York: McGraw-Hill.

[14] Nagasekhar and Tick-Hon Yip. (2005). "Finite Element Study of Multi pass Equal Channel Angular Extrusion/ Pressing." *Int. J. Nanosci.*, 4(4), 745-751.

[15] Nikulin, I., Kipelova, A., Malopheyev, S., & Kaibyshev, R. (2012). ""Effect of second phase particles on grain refinement during equal-channel angular pressing of an Al–Mg–Mn alloy." Acta Mater., 60(2), 487-497.

[16] Orlov, D., Todaka, Y., Umemoto, M., Beygelzimer, Y., Horita, Z., & Tsuji, N. (2009). "Plastic flow and grain refinement under simple shear-based

severe plastic deformation processing." *Mater. Sci. Forum*, 604, 171-178.

[17] Gajanan, M. N., Narendranath, S., & Kumar, S. S. (2019). "Effect of grain refinement on mechanical and corrosion behavior of AZ91 magnesium alloy processed by ECAE." *IOP Conf. Series: Mater. Sci. Eng.* 591(1), 012015.

[18] Naik, G.M., Narendranath, S., Satheesh Kumar, S.S. et al. Effect of Annealing and Aging Treatment on Pitting Corrosion Resistance of Fine-Grained Mg-8%Al-0.5%Zn Alloy. JOM 71, 4758-4768 (2019). https://doi.org/10.1007/s11837-019-03769-1.

[19] Agwa, M. A., Ali, M. N., & Al-Shorbagy, A. E. (2016). "Optimum processing parameters for equal channel angular pressing." *Mech. Mater.*, *100*, 1-11.

[20] Alaneme, K. K., & Okotete, E. A. (2019). "Recrystallization mechanisms and microstructure development in emerging metallic materials: A review." *J. Sci. Adv. Mater. Devices*, *4*(1), 19-33.

[21] Zhao, X., Li, S., Xue, Y., & Zhang, Z. (2019). "An Investigation on Microstructure, Texture and Mechanical Properties of AZ80 Mg Alloy Processed by Annular Channel Angular Extrusion." *Mater.,* 12(6), 1001.

[22] Qiao, J., Zheng, L., Ji, J., Bian, F., He, M., & Niu, T. (2018). "High temperature tensile behavior of Mg-2Al and Mg-6Al alloys." *Int. J. Mater. Res.*, 109(1), 28-33.

[23] Kim, W. J., Hong, S. I., Kim, Y. S., Min, S. H., Jeong, H. T., & Lee, J. D. (2003). "Texture development and its effect on mechanical properties of an AZ61 Mg alloy fabricated by equal channel angular pressing." *Acta Mater.*, 51(11), 3293-3307.

[24] Avvari, M., & Able, M. (2016). "Microstructure evolution in AZ61 alloy processed by equal channel angular pressing." *Adv. Mech. Eng., 8*(6), 1687814016651820.

[25] Avvari, M., Narendranath, S., & Nayaka, H. S. (2015). "A review on wrought magnesium alloys processed by equal channel angular pressing." *Int. J. Mater. Prod. Tech.*, *51*, 139-164.

[26] Avvari, M., & Narendranath, S. (2018). "Effect of secondary Mg 17 Al 12 phase on AZ80 alloy processed by equal channel angular pressing (ECAP)." *Silicon*, 10(1), 39-47.

[27] Avvari, M., & Narendranath, S. (2014). "Influence of Route-R on wrought magnesium AZ61 alloy mechanical properties through equal channel angular pressing." *J. Magnes. Alloys*, *2*(2), 159-164.

[28] Avvari, M., Narendranath, S., & Nayaka, H. S. (2014). "Effect of Processing Routes on AZ31 Alloy Processed By Severe Plastic Deformation." *Proc. Mater. Sci.*, *5*, 1560-1566.

[29] Yunchang, Tao Hu, and Paul K. Chu. (2010). "Influence of test solutions on in vitro studies of biomedical magnesium alloys." *J. Electrochemical Society* , 157 (7) , 238-243.

[30] Wenming, T. I. A. N., Songmei, L. I., Jianhua, L. I. U., Mei, Y. U., & Yujie, D. U. (2017). "Preparation of bimodal grain size 7075 aviation aluminum alloys and their corrosion properties." *Chinese J. Aero.*, 30(5), 1777-1788.

[31] Ambat, R., Aung, N. N., & Zhou, W. (2000). "Studies on the influence of chloride ion and pH on the corrosion and electrochemical behaviour of AZ91D magnesium alloy." *J. Appl. Electroch.*, *30*(7), 865-874.

[32] Shahar, I. A., Hosaka, T., Yoshihara, S., & Macdonald, B. J. (2017). "Mechanical and Corrosion Properties of AZ31 Mg Alloy Processed by

Equal-Channel Angular Pressing and Aging." *Proc. Eng.*, *184*, 423-431.

[33] Gopi, K. R., Nayaka, H. S., & Sahu, S. (2016). "Investigation of microstructure and mechanical properties of ECAP-processed AM series magnesium alloy." JMEP, *25*(9), 3737-3745.

[34] Lunder, O., Lein, J. E., Hesjevik, S. M., Aune, T. K., & Nişancioğlu, K. (1994). "Corrosion morphologies on magnesium alloy AZ 91." *Mater. Corr.*, *45*(6), 331-340.

[35] Singh, I. B., Singh, M., & Das, S. (2015). "A comparative corrosion behavior of Mg, AZ31 and AZ91 alloys in 3.5% NaCl solution." *J. Magnes. Alloys*, *3*(2), 142-148.

[36] Zeng, Rong-chang, Jin Zhang, Wei-jiu Huang, W. Dietzel, K. U. Kainer, C. Blawert, and K. E. Wei. (2006). "Review of studies on corrosion of magnesium alloys." *Trans. Nonferrous Met. Soc. China*, 16, 763-771.

[37] Cheng, Y. L., Qin, T. W., Wang, H. M., & Zhang, Z. (2009). "Comparison of corrosion behaviors of AZ31, AZ91, AM60 and ZK60 magnesium alloys." *Trans. nonferrous met. Soc. China, 19*(3), 517-524.

Chapter 2

Magnesium Containing High Entropy Alloys

Prince Sharma, Nushrat Naushin, Sahil Rohila and Abhishek Tiwari

Abstract

High Entropy alloys (HEAs) or Complex Concentrated Alloys (CCAs) or Multi-Principal Element Alloys (MPEAs) is a matter of interest to material scientists for the last two decades due to the excellent mechanical properties, oxidation and corrosion resistant behaviors. One of the major drawbacks of HEAs is their high density. Mg containing HEAs show low density compared to peers, although extensive research is required in this field. This chapter aims to include all the available information on synthesis, design, microstructures and mechanical properties of Mg containing HEAs and to highlight the contemporary voids that are to be filled in near future.

Keywords: Magnesium, High Entropy Alloys, Light Weight Alloys, Microstructure, Mechanical Properties

1. Introduction

The ancient strategy of alloy design and production has been followed for a long period and it will remain a crucial part of industry. The strategy is based on one principal element solvent with solute elements dissolved in the lattice either as heterogeneities or precipitate particles. Although, ancient Indian scriptures in Sanskrit mentions the existence of multi principal alloys named as "Tri-loh" (loh means Iron), "Panch-dhatu" and "Asht-dhatu" in which Tri, Panch and Asht mean three, five and eight respectively [1]. These dhatus or metals were used for special and sacred purposes such as making idols, deities and machines [2]. The context of these alloys can be found in books and Sanskrit text titled "Vimanika Shastra", "Ras-Ratnakar-Samuchaya", "Shilparatna", "Manasara", "Ras-Tarangini", and "Ras-grandhas". These alloys were used for special purposes and the knowledge of their synthesis was mostly limited to the *Sages* and *Rishis*, which has never been researched and explored by the scientific community. The concept of multi principal element or high entropy alloy is modern version of the "Panch-dhatu".

The concept of maximization of configurational entropy by using five or more elements in nearly equi-atomic composition has revolutionized the field of alloy design and metallurgy. The first scientific report of such alloys were reported independently by Cantor et. al. and Yeh et. al [3, 4]. HEAs show better mechanical, oxidation, corrosion and irradiation properties compared to commercial alloys. HEAs show four key effects and a postulate, which are: high-entropy effect, severe lattice distortion, sluggish diffusion, short range order effect and cocktail effect [5].

2. Scope of this chapter

The authors aimed to present detailed review and analysis of design, synthesis, microstructures and mechanical properties of Mg containing MPEAs from all the existing scientific articles on the topic available to this date. The presence of multiple principal elements in an alloy makes it difficult to interpret the reasons behind the behavior of HEAs. This study presents the possible interpretations of the structure-properties-processing relationship of Mg containing alloys in both equiatomic and non-equiatomic compositions. **Figure 1** shows the gradual increase in interest in Mg-HEAs.

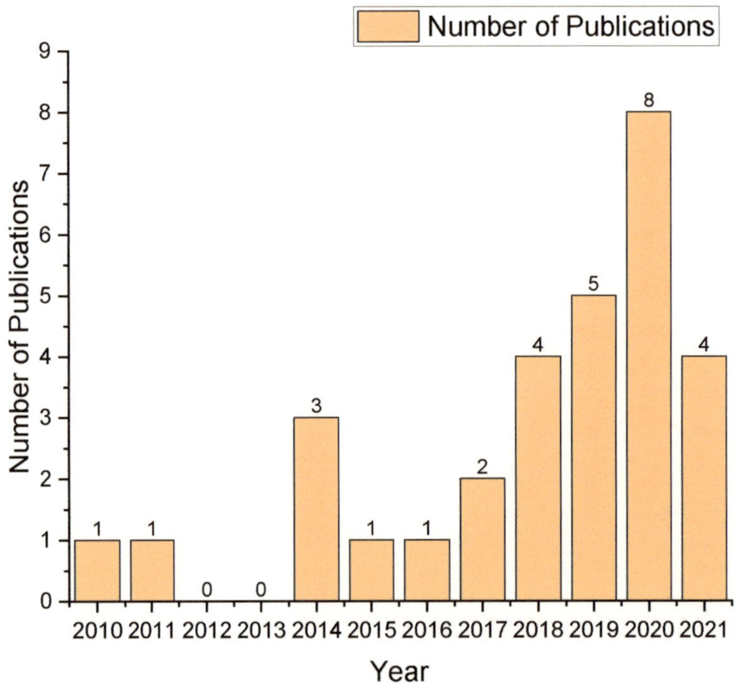

Figure 1.
Year vs number of publications of Mg containing HEAs.

3. Basics of HEAs

HEAs are multicomponent systems containing five or more elements in significant proportions unlike conventional alloys [6]. This leads to increase in the overall configurational entropy of system, which was first reported by Yeh et. al. for CuCoNiCrAlFe system [4]. It was believed that due to high configurational entropy of mixture, the alloys tend to crystallize as SS instead of intermetallic compounds (IM). These compositions exhibit superior mechanical, oxidation and corrosion resistance properties under conditions ranging from high temperatures to cryogenic temperatures [5, 6]. Due to the prominence of high entropy effect in multi-principal element alloys, such alloys are commonly referred as high entropy alloys. There are four core effects and one postulate, as follows:

3.1 The high entropy effect

The name perhaps has gathered enormous attention of researchers in this system. It states that high configurational entropy of mixing (ΔS_{conf}) supports the formation of SS instead of IM [4]. This also limits the number of phases predicted by the Gibbs Phase Rule. ΔS_{conf} is given by Eq. (1),

$$\Delta S_{conf} = -R\sum_{i=1}^{n} X_i \ln X_i \tag{1}$$

On dealing with this phenomenon on the grounds of probability, the possible explanation of absence of IM is in the basic nature of these compounds. Intermetallics are strictly ordered in nature but in case of HEAs, there is interaction with various elements with significantly higher compositions. As a result, even if thermodynamic and kinetic factors favor compound formation, the probability of tendency to form a two elements compound being in vicinity, is reduced. Intermetallics (IMs) are established by a high value of negative enthalpy of mixing (ΔH_{mix}) and so entropy maximization plays an important role in lowering the Gibbs free energy of mixing and promoting SS formation. Here, ΔH_{mix} Denotes enthalpy of mixing of all elements and is given by Eq. (2). In cases, where ΔH_{mix} of an intermetallic is higher compared to the solid solution phases, there is formation of IM and SS together. The Ω factor introduced by Yang and Zhang [7], should ideally be greater than 1.1 to stabilize the SS.

$$\Delta H_{mix} = \sum_{i=1, j>1}^{n} 4\Delta H_{AB}^{mix} c_i c_j \tag{2}$$

$$\Omega = \frac{T_m \Delta S_{conf}}{|\Delta H_{mix}|} \tag{3}$$

$$T_m = \sum_{i=1}^{n} T_i c_i$$

3.2 The lattice distortion effect

Unlike conventional alloys with a single element solvent matrix, the HEAs comprises of a random solute matrix (RSM). The assumption, that RSM comprises of constituting elements arranged with complete disordered fashion and hence result in a SS, is mostly true. There is occasional occurrence of some intermetallic phases in the alloy which could be due to (a) enthalpy of formation of intermetallic is very high compared to the enthalpy of mixture; (b) the difference in atomic size is less (1.1-1.6) which supports formation of ordered structure; (c) the incorporation of elements with very small atoms in the alloy, which forms interstitial compounds. The Lattice Distortion Effect (LDE) is theoretically estimated by polydisparity or atomic mismatch factor δ given by Eq. (4) [7].

$$\delta = \sqrt{\sum\nolimits_{i=1}^{n} c_i \left(1 - \frac{r_i}{\sum\nolimits_{j=1}^{n} c_j r_j}\right)^2} \tag{4}$$

The LDE tends to increase the hardness of HEAs because of solution hardening.

3.3 The sluggish diffusion hypothesis

Studies suggest that MPEAs exhibit a diffusion rate slower than any regular alloys and which could be responsible for the lower number of phases present in MPEAs [8]. This slow diffusion behaviour is claimed as sluggish diffusion. The HEAs studied at the beginning exhibited this behavior, though the sluggish

Sr. No.	HEA Composition	Mg	Phases	Tm (K)	ΔS/R	VEC	δ	Δχ	ΔH (KJmol^{-1})	Ω	Density
1	${}^{Al}8^{Li}0.5^{Mg}0.5^{Sn}0.5^{Zn}0.5$	0.05	SS + IM	875.55	0.78	3.35	0.0398	0.17	−0.54	10.48	3.05
2	${}^{Al}8^{Cu}0.5^{Li}0.5^{Mg}0.5^{Zn}0.5$	0.05	SS + IM	918.18	0.78	3.70	0.0424	0.17	−1.15	5.16	3.08
3	$Al_{60}Cu_{10}Fe_{10}Cr_5Mn_5Ni_5Mg_5$	0.05	SS + IM	1191.78	1.37	4.95	0.0736	0.14	−7.91	1.71	4.6
4	$Al_{35}Cr_{14}Mg_6Ti_{35}V_{10}$	0.06	SS	1281.16	1.13	3.07	0.1561	0.22	−15.49	0.78	4.05
5	${}^{Al}20^{Li}20^{Mg}10^{Sc}20^{Ti}30$	0.10	SS	1315.04	1.56	2.80	0.0455	0.23	−0.40	42.56	2.67
6	${}^{Mg}0.10^{Ti}0.30^{V}0.25^{Zr}0.10^{Nb}0.25$	0.10	SS + IM	2208.40	1.56	4.74	0.0722	0.12	6.08	4.71	NA
7	$AlFeCuCrMg_{0.5}$	0.11	SS	1487.12	1.58	6.44	0.08	0.18	4.05	4.83	5.79
8	$MgMoNbFeTi_2$	0.17	SS	2042.21	1.56	4.83	0.0662	0.27	6.22	4.26	NA
9	$MgMoNbFeTi_2Y_{0.004}$	0.17	SS	2041.98	1.56	4.83	0.0662	0.27	6.22	4.26	NA
10	$MgMoNbFeTi_2Y_{0.008}$	0.17	SS	2041.99	1.56	4.83	0.0662	0.27	6.22	4.26	NA
11	$MgMoNbFeTi_2Y_{0.012}$	0.17	SS	2042.00	1.56	4.83	0.0662	0.27	6.22	4.26	NA
12	AlLIMgSnZn	0.20	SS + IM	701.63	1.61	4.40	0.0611	0.33	−6.24	1.50	4.23
13	AlFeCuCrMg	0.20	SS + IM	1430.84	1.61	6.00	0.0925	0.21	6.24	3.07	5.37
14	$Mg_{20}(MnAlZnCu)_{80}$	0.20	SS + IM	1084.80	1.61	7.00	0.117	0.19	−3.04	4.77	4.29
15	MgVAlCrNi	0.20	SS	1154.90	1.29	4.00	0.2335	0.35	−6.24	1.98	NA
16	MgAlSiCrFe	0.20	SS + IM	1159.30	1.29	3.80	0.2473	0.33	2.88	4.31	NA
17	Al2MgLiCa	0.20	IM	873.35	1.33	2.20	0.1302	0.28	−9.44	1.02	NA
18	MgVAlCr	0.25	SS	1544.13	1.39	4.00	0.0832	0.14	4.25	4.19	NA
19	MgVAlNi	0.25	SS	1443.63	1.39	5.00	0.0602	0.21	−9.75	1.71	NA
20	MgVCrNi	0.25	SS	1742.75	1.39	5.75	0.0861	0.21	4.00	5.02	NA
21	AlMgLiCa	0.25	IM	858.30	1.39	2.00	0.1285	0.26	−8.25	1.20	NA
22	$Mg_{26}V_{31}Al_{31}Cr_6Ni_6$	0.26	SS	1439.75	1.41	3.96	0.0703	0.16	−2.10	8.05	NA

Sr. No.	HEA Composition	Mg	Phases	Tm (K)	ΔS/R	VEC	δ	Δχ	ΔH (KJmol^{-1})	Ω	Density
23	$Mg_{28}V_{28}Al_{19}Cr_{19}Ni_6$	0.28	SS	1557.39	1.51	4.27	0.0826	0.17	3.82	5.13	NA
24	$AlFeCuCrMg_{1.7}$	0.30	SS + IM	1368.19	1.58	5.51	0.1	0.22	7.99	2.25	4.91
25	$AlMgLiCa_{0.3}$	0.30	IM	802.19	1.30	2.00	0.0962	0.26	−5.18	1.68	NA
26	$AlLi_{0.5}MgSn_{0.2}Zn_{0.5}$	0.31	SS + IM	790.87	1.48	3.84	0.061	0.27	−3.98	2.44	3.22
27	$AlCu_{0.2}Li_{0.5}MgZn_{0.5}$	0.31	IM	844.16	1.48	4.28	0.0702	0.26	−3.40	3.06	3.73
28	$AlCu_{0.5}Li_{0.5}MgSn_{0.2}$	0.31	SS + IM	894.76	1.48	3.69	0.0774	0.31	−3.65	3.02	3.69
29	$Mg_{33}(MnAlZnCu)_{67}$	0.33	SS + IM	1058.51	1.56	8.03	0.2294	0.45	−3.69	3.72	3.41
30	$Mg_{35}Al_{33}Li_{15}Zn_7Ca_5Cu_5$	0.35	SS + IM	893.96	1.50	2.93	0.1339	0.24	−7.44	1.50	2.25
31	$Mg_{35}Al_{33}Li_{15}Zn_7Ca_5Cu_5$	0.35	SS + IM	871.70	1.50	3.33	0.1091	0.26	−4.97	2.19	2.27
32	$Mg_{43}(MnAlZnCu)_{57}$	0.43	SS + IM	1038.29	1.47	5.56	0.1209	0.21	−1.38	9.22	2.71
33	$Mg_{45.6}(MnAlZnCu)_{54.4}$	0.46	SS + IM	1033.03	1.44	5.40	0.12	0.21	−1.23	10.07	2.53
34	$Mg_{50}(MnAlZnCu)_{50}$	0.50	SS + IM	1024.13	1.39	5.13	0.1181	0.21	−1.00	11.80	2.2
35	$Mg_{80}Al_5Cu_5Mn_5Zn_5$	0.80	SS + IM	963.45	0.78	3.25	0.085	0.16	−0.04	155.73	1.74

Table 1
Value of empirical parameters of Mg containing HEAs.

diffusion is not entirely common for every HEA. Depending upon the alloy composition, the behavior varies [9]. The diffusion studies have been conducted in a limited number of alloys. Earlier studies on CoCrFeMnNi at near T_m temperatures exhibit sluggish diffusion as per diffusion couple method and quasi binary path [10, 11]. A few studies suggest that the diffusion is not sluggish in CoCrFeMnNi within 1073K-1373K temperature range, when analyzed with Radiotracer method using ^{63}Ni. Bulk diffusion is the prominent mode of diffusion in a single crystal CoCrFeMnNi, while bulk diffusion is also in polycrystalline alloys at high temperatures [9]. Using Boltzman-Matano method to study diffusion behaviour also reveals similar results [12]. Radiotracer method using ^{51}Cr, ^{54}Mn, ^{57}Cr, ^{59}Fe and ^{63}Ni in the same alloy also produced similar non-sluggish diffusion behaviour when analyzed at a temperature range of 1073K-1373K [13–17]. AlCoCrFeNi HEAs exhibit sluggish diffusion behaviour when analyzed using diffusion couple method [18]. Numerical assessments also suggest similar behaviour. Another study on $Al_xCoCrCuFeNi$ where x=1, 1.5, 1.8 suggests that the diffusion retards with increasing the Al-concentration [19].

3.4 The Cocktail effect

Cocktail effect is a postulate, first identified by Prof. S. Ranganathan [20]. It refers to the unpredictable improvement in properties of materials based on the synergy of multi principal alloying elements. The properties range from near zero thermal expansion, ultra-high strength, corrosion resistance, good fracture toughness, ductility to photovoltaic effects or thermo-electronic responses [5]. This effect reminds about the acceptance of unique properties formed due to unusual combinations of elements as observed in MPEAs.

3.5 The short-range ordering

Besides these effects, chemical short range ordering (CSRO) has a significant impact on the mechanical properties, as it is proved to facilitate phase transformations in HEAs [21]. The CSRO has direct correlation with the activation energy of phase transformation from FCC to HCP phase in CoCrNi alloys [22]. The presence of SRO increases the Stacking Fault Energy (SFE) and yield strength in several alloys; it has been proved for CoCrNi alloys both experimentally and computationally [21–23]. A study using reverse Monte-Carlo suggests that the presence of SRO in TiVNb alloys is responsible for forming the FCC supercells in the alloy [21, 24].

All the empirical parameters discussed in above sections are calculated on the basis of available information in **Table 1**.

4. Why Mg HEAs

Since the inception, there have been several doubts, questions and strong arguments against HEAs. Most of them are related to scalability and process engineering in industries, repeatability, reliability, applications and high density of HEAs. Among all these drawbacks, process engineering of HEAs is least studied, while the major problem are repeatability and their high density. To obtain a high level of repeatability, an extensive standard of alloy preparation and their further processing must be established throughout the world. The problem of high density is being solved with the development of light weight HEAs (LHEA) containing low density elements such as Al, Ti, Li and Mg. LHEAs consisting of Mg and Li are found to be lightest [25]. Mg based alloys have several applications in aircraft and automobile panels, bio-implants and energy storage. Mg alloys show exceptional and beautiful microstructures with Long Period Stacking Order (LPSO) phases. LPSO phases are most important feature of few Mg based alloys systems Mg-TM-RE (TM: transition metal, RE: rare earth metal) alloy as it enhances the mechanical properties at room and elevated temperatures [26]. ***LPSO phases are not yet reported in Mg-HEAs and could be one of the most interesting breakthrough in Light-weight High Entropy Alloys (LHEAs)***.

There are several other factors which makes Mg-HEAs a topic of interest for example, Magnesium alloy anodes tends to increase the efficiency of Mg-ion batteries which may replace Li-ion batteries in future; Mg is a fast biodegradable and bio-compatible material. Magnesium alloys do not possess enough strength compared to the bone tissue. Mg containing HEAs could be the answer to this problem. Lynette W. Cheah found that for every 10 % mass reduction in vehicle, fuel consumption may reduce by 7 % [27]. If the parts of vehicle are made of strong alloys containing Mg and/or Al instead of Iron, the weight reduction will be 45 and 29 % respectively. This will significantly lower the carbon emission and save fossil fuel. The disadvantage with high reactivity of Mg and low strength of its alloys can be avoided by the virtue of introducing severe lattice distortion, high entropy effect and cocktail effect. This means that Mg must be alloyed with three or more elements to increase the configurational entropy of alloy to attain higher stability and strength.

Clearly, Mg containing HEAs are materials for future.

Alloying has a positive impact on the properties of Mg, and it has been proven that Al addition increases hardness, strength and castability without significantly affecting the density [28]. Ca enhances the thermo-mechanical properties, increases creep resistance and refines the grains. Nd and Ni both increase the strength of Mg when added separately. Cu enhances mechanical properties and aid thermal stability. Ce addition improves corrosion resistance; Mn increases saltwater corrosion resistance in Mg-Al alloys; Zn increases corrosion resistance in Mg-Ni-Fe alloys; Sn prevents cracks during Mg-Al alloy processing and Sr increases creep resistance [28, 29]. Every element has a unique effect on alloy, as shown in **Figure 2** and hence, HEAs must be exploited to establish a synergy between different alloying elements in Mg to produce an alloy with high strength to weigh ratio. In **Figure 2**,

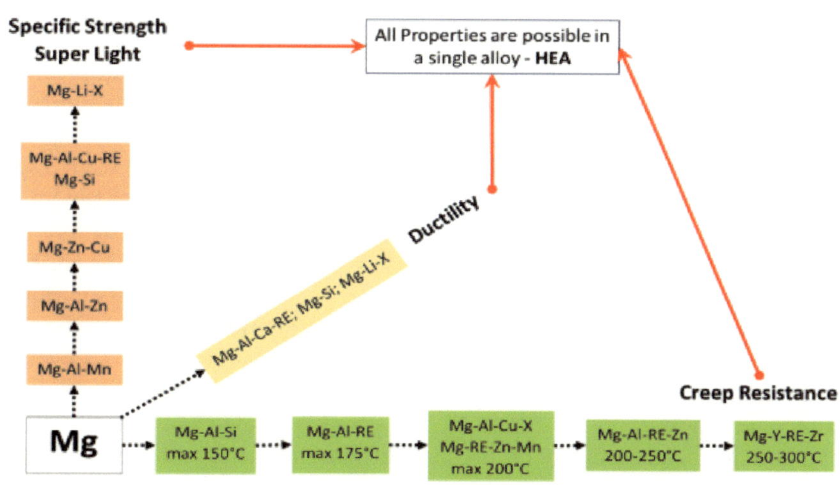

Figure 2.
Development of magnesium alloys.

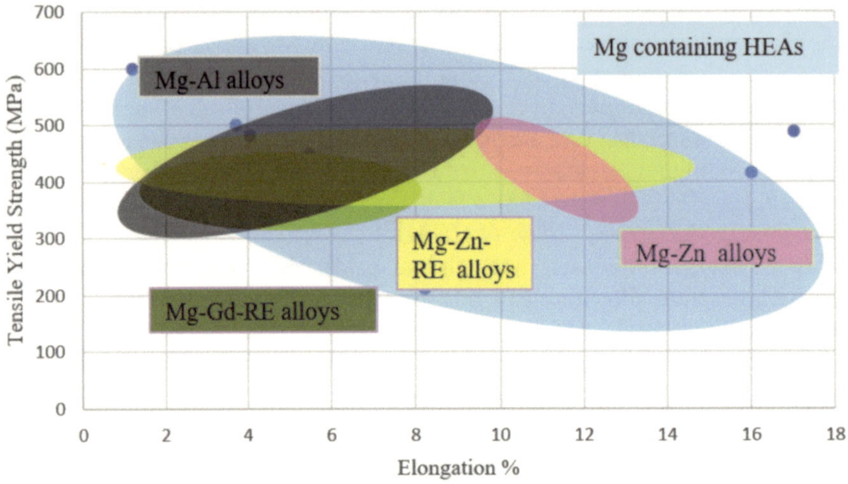

Figure 3.
Tensile yield strength versus elongation %.

Mg containing HEAs may have property in a combination of all (cocktail effect). Most Mg containing HEAs show light weight and moderate strength which is described in the mechanical properties section 6. A combination of the light weight achieved due to Mg being one of the base metals and the superior properties of the other principal elements makes HEAs special.

Figure 3 shows graph of tensile yield strength versus elongation of a range of Mg-Al, Mg-Zn, Mg-Zn-RE, Mg-Gd-RE alloys and Mg containing HEAs [30–35]. The available data on Mg containing HEAs are used in this plot and the other Mg alloy range is obtained from a study by Sankaran and coworkers [30]. This diagram shows the wide range of tensile strength and elongation depending on the compositions of alloys, which not only contributes to the materials property but also adds new possible alloys for a wide range of applications.

5. Synthesis of Mg HEA

Figure 4 shows Al is the highest alloyed element with Mg followed by Cu and Li in HEA system. Mg is a reactive metal with a low melting point. It burns with a shiny white light in air. Considering its high vapour pressure, Mechanical Alloying (MA) would be a suitable process compared to melting and casting for synthesis of alloys containing Mg. MA is a common route for synthesis of Mg containing HEAs [33, 36–42]. Youssef et. al [37] and Ornov et. al [43] were the first to study Mg containing HEAs synthesized by MA. Initial results were promising for the future of Mg-HEAs. $AlFeCuCrMg_x$ (x = 0, 0.5, 1, 1.7 mol) was synthesized by MA process to understand the effect of composition of Mg on phases and solid solution formability. It has been found that Mg addition stabilizes BCC structure via dissolution of FCC structure [43]. Cold welding supersedes fracturing in alloys with higher Mg. Lattice distortion was found to increase with Mg % in the alloys [43]. A low density Mg containing HEA with FCC phase transformed to more stable HCP structure on annealing and reported high hardness of 5.8 GPa [37]. Due to continuous cold welding and mechanical impact by the balls the atomic density of the powder increases and hence number of slip systems and ductility increases. So, the milling time must be increased with the increase in Mg composition [43]. Maulik and Kumar confirmed lattice distortion, which increased with further addition of Mg

[43]. Although, for sintered alloys, lattice distortion first increased and then decreased with Mg content [38]. Equiatomic MgAlSiCrFe HEA was synthesized by MA and was found to be stable at moderate temperatures (300-500°C), also Si did not dissolve completely even after 60 hours of milling [44]. In the pioneer works on Mg containing HEAs, a similar phenomenon could be identified that conventional threshold limits of ΔH_{mix} and δ do not support the formation of solid solution but there is still formation of SS [6, 7, 35, 43, 45].

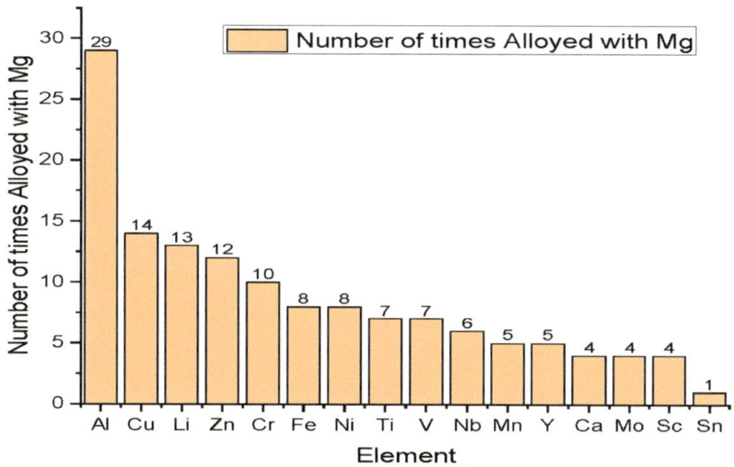

Figure 4.
Elements vs number of times alloyed with Mg.

Element	Symbol	Atomic Mass	r (Å)	VEC	X	Density	Tm (K)
Lithium	Li	6.941	1.57	1	0.98	0.534	453.69
Magnesium	Mg	24.305	1.6	2	1.31	1.738	923
Aluminium	Al	26.9815	1.43	3	1.61	2.7	933.52
Calcium	Ca	40.078	2	2	1	1.55	1123
Scandium	Sc	44.9559	1.6	3	1.36	2.985	1812
Titanium	Ti	47.88	1.47	4	1.54	4.506	1943
Vanadium	V	50.9415	1.36	5	1.63	6.11	2190
Chromium	Cr	51.996	1.28	6	1.66	7.19	2130
Manganese	Mn	54.93805	1.12	7	1.55	7.21	1517
Iron	Fe	55.847181	1.28	8	1.83	7.874	1810
Nickel	Ni	58.69	1.5	10	1.91	8.908	1728
Copper	Cu	63.546	1.28	11	1.9	8.96	1357.7
Zinc	Zn	65.39	1.37	12	1.65	7.14	692.8
Yttrium	Y	88.9059	1.81	3	1.22	4.472	1803
Niobium	Nb	92.9064	1.47	5	1.6	8.57	2750
Molybdenum	Mo	65.9064	1.4	6	2.16	10.28	2883
Tin	Sn	118.71	1.58	4	1.96	7.265	505.12

Table 2
Physical properties of elements.

The ΔH_{mix} values in **Table 2** are taken from [46, 47] in which the values have been calculated using Miedema Model [48–52]. A significant portion of the high entropy alloys are produced using the induction melting route [32, 34, 35]. It can be easily inferred from the **Table 2** that there is unavoidable formation of intermetallic and Al-Mn quasicrystals which is majorly because of high negative values of ΔH_{mix} for various binary pairs given in **Table 2**.

6. Microstructure and mechanical behavior of magnesium containing HEAs

Microstructure of a material has a direct influence on its mechanical behaviors. Equiatomic MgMnAlZnCu HEA produced using induction melting consists of a matrix and a floral pattern where the floral pattern is rich in Al-Mn icosahedral quasicrystals and the matrix consists of an HCP phase comprised of all the alloying elements [31]. The presence of the quasicrystals is responsible for the increased hardness in this alloy, which gradually increases upon increasing the cooling rate. It is worth noting that Al-Mn quasicrystals are thermally stable. The increase in cooling rate also changes the plasticity of the alloy and hence, it increases the elasticity [31]. The microstructures are given in the following **Figure 5(a)**. Increasing the amount of Mg in the alloy reduced the ΔH_{mix} , as a result, there is increase in the amount of SS phases [32]. Along with the HCP and icosahedral quasicrystalline phases, a pure Mg and Mg_7Zn_3 phases are present in Magnesium

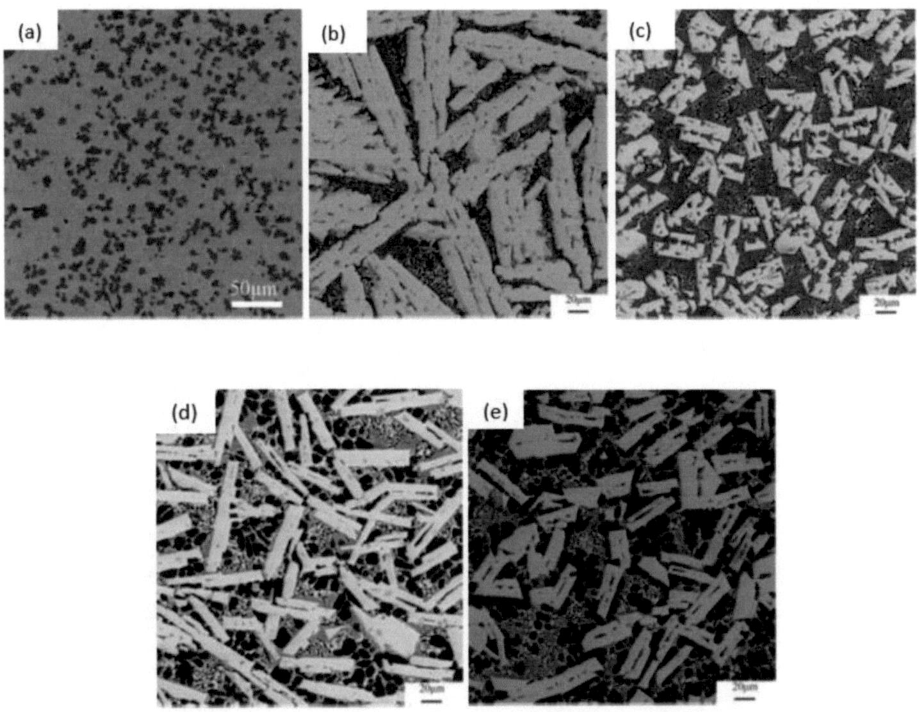

Figure 5.
(a) Equiatomic MgMnAlZnCu (Induction melting in Ar atmosphere, cooled using brine in Cu mold), (b) $Mg_{33}(MnAlZnCu)_{67}$ *(Induction melting in Ar atmosphere, Air cooling in Cu mold), (c)* $Mg_{43}(MnAlZnCu)_{57}$ *(Induction melting in Ar atmosphere, Air cooling in Cu mold), (d)* $Mg_{45.6}(MnAlZnCu)_{54.4}$ *(Induction melting in Ar atmosphere, Air cooling in Cu mold), (e)* $Mg_{50}(MnAlZnCu)_{50}$ *(Induction melting in Ar atmosphere, Air cooling in Cu mold) (Reprinted with permission from Refs. [31,32]).*

containing entropy stabilized alloy systems. Increase in Mg content enhances the complexity of the microstructures as per the **Figure 5(b)-(e)** [32].

Another light weight Mg-HEA, $Al_{60}Cu_{10}Fe_{10}Cr_5Mn_5Ni_5Mg_5$, fabricated using vacuum induction melting followed by die casting, exhibits three different phases (1) Al_3Fe_4, (2) Al_7Cu_4Ni and (3) $Mg_2Cu_6Al_5$ in the SEM image (**Figure 6**) [34].

Tun et. al. fabricated $Mg_{80}Al_5Cu_5Mn_5Zn_5$ HEA through disintegrated melt deposition followed by extrusion: The alloy showed two IM and one SS phase in the microstructure [53]. The microstructure contained Al_6Mg and Al_2CuMg intermetallics and an HCP phase containing Mg, Mn and Al. The microstructure of $Mg_{80}Al_5Cu_5Mn_5Zn_5$ HEA is shown in **Figure 7(a)**. Phases mentioned as 1, 2 and, 3 are the HCP phase, Al_6Mg and Al_2CuMg phases respectively.

The density of any high entropy system is highly dependent on the elements present in the alloy. Mg and Li system produces extremely light weight alloys while addition of Al increases density but it strengthens the system as well. This increases the specific strength, which makes magnesium based high and medium entropy systems a matter of interest these days. The highest density in any magnesium containing multi-element alloy system is 5.06 g/cc in equiatomic MgMnAlZnCu produced using induction melting followed by cooling in brine, whereas $Mg_{80}Al_5Cu_5Mn_5Zn_5$ exhibits the lowest density of 2.15 g/cc. The other available density data is given in **Table 3**.

The hardness of Mg-HEAs generally decreased with increase in composition of Mg, the maximum hardness with equiatomic composition was found to be 428 HV [32]. MgMnAlZnCu HEA exhibits the highest hardness when processed at higher cooling rate due to formation of Al-Mn icosahedral quasicrystals [31].

Mg-HEAs with a low magnesium content exhibit higher yield strength. The $Al_{60}Cu_{10}Fe_{10}Cr_5Mn_5Ni_5Mg_5$ exhibits a yield strength of 743 MPa, while magnesium rich systems can result in yield stress as low as 211 MPa. The yield strength data

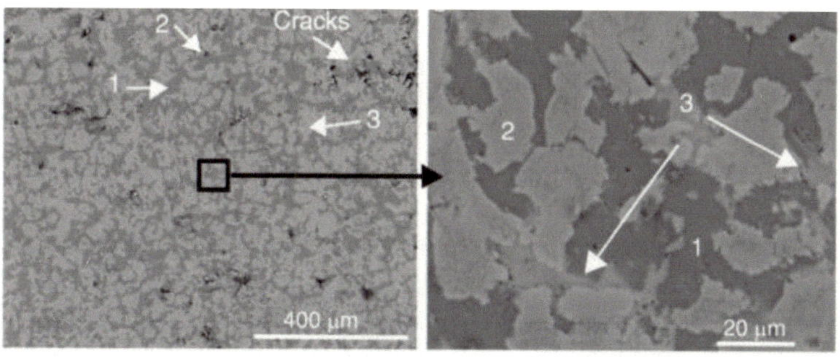

Figure 6.
SEM micrographs of $Al_{60}Cu_{10}Fe_{10}Cr_5Mn_5Ni_5Mg_5$ *(Reprinted with permission from Ref. [34]).*

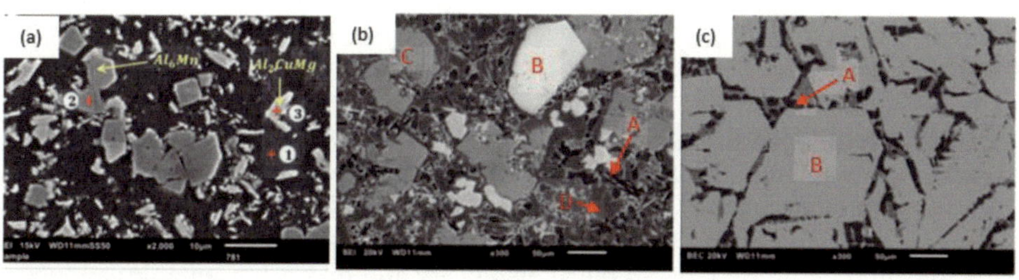

Figure 7.
SEM micrograph of (a) $Mg_{80}Al_5Cu_5Mn_5Zn_5$, *(b)* $Mg_{35}Al_{33}Li_{15}Zn_7Ca_5Y_5$ *and (c)* $Mg_{35}Al_{33}Li_{15}Zn_7Ca_5Cu_5$ *(image (a) adapted from Ref. [53], image (b) and (c) Reprinted with permission from Ref. [33]).*

	Mg	Al	Cu	Li	Zn	Fe	Ti	Cr	Mn	Nb	V	Ni	Mo	Sn	Sc	Ca	Y
Al	-2																
Cu	-3	-1															
Li	0	-4	-5														
Zn	-4	1	1	-8													
Fe	18	-11	13	26	4												
Ti	16	-30	-9	34	-15	-17											
Cr	24	-10	12	35	5	-1	-7										
Mn	10	-19	4	19	-6	0	-8	2									
Nb	32	-18	3	-46	-1	-16	2	-7	-4								
V	23	-16	5	37	-2	-7	-2	-2	-1	-1							
Ni	-4	-22	4	1	-9	-2	-35	-7	-8	-30	-18						
Mo	36	-5	19	49	12	-2	-4	0	5	-6	0	-7					
Sn	-9	4	7	-18	1	11	-21	10	-7	-1	-1	-4	20				
Sc	-3	-38	-24	12	-29	-11	8	1	-8	18	7	-39	11	-45			
Ca	-6	-20	-13	-1	-22	25	43	38	19	63	44	-7	56	-45	17		
Y	-6	-38	-22	8	-31	-1	15	11	-1	30	17	-31	24	-51	1	11	

Table 3
Enthalpy of mixing of elements in a pair.

testifies for the brittleness caused due to presence of high amount of Mg. The behavior is predominantly because of HCP structure induced by higher percentage of Mg which inhibits dislocation by insufficient number of slip systems. Although by adding Li with 10.3 wt% in Mg the crystal structure changes from HCP to BCC. The properties of Mg-HEAs are better than conventional Magnesium alloys. The reason behind this lies in the formation of a complex mixture of intermetallics and solid-solution phases in the system according to a study on $Mg_{80}Al_5Cu_5Mn_5Zn_5$ by Tun et. al. [53].

It can be inferred from existing literature, Mg-HEAs synthesized by MA can find applications in structural as well as hydrogen storage systems [39, 41, 54–57] although more extensive research is required to find a better alloy and processing techniques; whereas induction melting in different atmospheres and casting process were used for alloys to be used in load bearing components. Mechanical properties are available for a few alloys which makes it difficult for authors to give a comprehensive analysis. $MgMoNbFeTi_2$ and Y doped $MgMoNbFeTi_2$ were synthesised using mechanical alloying followed by laser cladding. Increase in Yttria content resulted in enhanced mechanical properties of the alloy [36]. It is well established that for enhancement of mechanical properties, Mg alloys are doped with reactive elements like Re, Y and Hf.

Table 4 represents the manufacturing route for various Mg containing HEAs, the phases present in corresponding alloys, the presence of intermetallic phases and the physical and mechanical properties of the alloys. It is evident from the data that the presence of intermetallics enhance the mechanical properties of the alloys. The increased amount of Mg reduces the alloy density and makes it a suitable candidate for aerospace applications. Further researches on Mg-HEAs need to aim on creating light weight and strong alloys at the same time, this could be done by *thermo-mechanical processing*, which is not yet explored for Mg-HEAs.

HEA Composition	Processing Route	Phases				Reported Properties				Year	Ref
		FCC	BCC	HCP	Intermetallics	Tensile Strength (MPa)	Compressive Strength (MPa)	Hard-ness (HV)	Density (g/cc)		
$Al_{20}Li_{20}Mg_{10}Sc_{20}Ti_{30}$	Mechanical alloying			✓					3.05	2015	[37]
AlLiMgSnZn	Induction Melting	✓		✓	✓		615		3.88	2014	[35]
$AlLi_{0.5}MgSc_{0.2}Zn_{0.5}$		✓			✓		546		2.9		
$AlCu_{0.2}Li_{0.5}MgZn_{0.5}$					✓				2.75		
$AlCu_{0.5}Li_{0.5}MgSn_{0.2}$					✓				2.96		
$Al_8Li_{0.5}Mg_{0.5}Sn_{0.5}Zn_{0.5}$		✓			✓		836		3.05		
$Al_8Cu_{0.5}Li_{0.5}Mg_{0.5}Zn_{0.5}$		✓			✓		879		2.91		
$AlFeCuCrMg_{0.5}$	Mechanical Alloying		0.44		0.56				NA	2017	[38]
AlFeCuCrMg			0.505		0.495				NA		
$AlFeCuCrMg_{1.7}$			0.1		0.9 (Cu_2Mg major)				NA		
$Mg_{20}(MnAlZnCu)_{80}$	Induction Melting			✓	Al-Mn quasicrystal		428	428	4.3	2010	[32]
$Mg_{33}(MnAlZnCu)_{67}$				✓	Al-Mn quasicrystal		437	324	3.5		
$Mg_{43}(MnAlZnCu)_{57}$				✓	Al-Mn quasicrystal		500	244	2.5		
$Mg_{45.6}(MnAlZnCu)_{54.4}$				✓	Al-Mn quasicrystal		482	223.2	2.3		
$Mg_{50}(MnAlZnCu)_{50}$				✓	Al-Mn quasicrystal		400	178	2.2		
$Mg_{80}Al_5Cu_5Mn_5Zn_5$	Disintegrated melt deposition.			✓	✓	318	616	196	2.15	2019	[53]
MgVAlCr	Mechanical Alloying		✓						NA	2021	[39]
MgVAlNi			✓						NA		
MgVCrNi			✓						NA		
MgVAlCrNi			✓						NA		
$Mg_{28}V_{28}Al_{19}Cr_{19}Ni_6$			✓						NA		
$Mg_{26}V_{31}Al_{31}Cr_6Ni_6$			✓						NA		

HEA Composition	Processing Route	Phases				Reported Properties				Year	Ref
		FCC	BCC	HCP	Intermetallics	Tensile Strength (MPa)	Compressive Strength (MPa)	Hard-ness (HV)	Density (g/cc)		
$Mg_{0.1}Ti_{0.3}V_{0.25}Zr_{0.1}Nb_{0.25}$	Mechanical Alloying		✓						NA	2021	[41]
$Al_{35}Cr_{14}Mg_{6}Ti_{33}V_{10}$	Mechanical Alloying		✓	✓		1503		460	4.91	2019	[42]
MgAlSiCrFe	Mechanical Alloying	minor	✓		Mg_2 Si (Major)				NA	2020	[44]
AlMgLiCa	Casting							NA	NA	2020	[58]
$Al_2MgLiCa$								≈300	NA		
$AlMgLiCa_{0.3}$								NA	NA		
$Al_{60}Cu_{10}Fe_{10}Cr_5Mn_5Ni_5Mg_5$	Vacuum Induction Melting	✓			✓			743	NA	2018	[34]
$MgMoNbFeTi_2$	Mechnaical Alloying;laser cladding (Coating)		✓					≈400	NA	2020	[36]
$MgMoNbFeTi_2Y_{0.004}$			✓					≈500 HV	NA	2020	
$MgMoNbFeTi_2Y_{0.008}$		✓	✓					≈650	NA	2020	
$MgMoNbFeTi_2Y_{0.012}$		✓	✓					≈1000	NA	2020	
$Mg_{35}Al_{33}Li_{15}Zn_7Ca_5Y_5$	Disintegrated melt deposition			α-Mg	✓			237 ± 10	2.25	2018	[33]
$Mg_{35}Al_{33}Li_{15}Zn_7Ca_5Cu_5$				α-Mg	✓			267 ± 15	2.27	2018	

Table 4.
Manufacturing routes, present phases and corresponding mechanical properties of Mg containing HEAs.

7. Computational approach in phase identification of magnesium containing high entropy alloys

A large number of researches have been conducted on binary and ternary alloys but quaternary and quinary systems are yet to be explored. The number of possible HEAs are around 10^{177} [59]. In order to screen out alloys for use, Calculation of Phase Diagrams (CALPHAD) is an exceptional tool. CALPHAD can be time consuming compared to Molecular Dynamics (MD) and Density Functional Theory (DFT) simulations. CALPHAD uses the phase diagrams and thermodynamics to determine the phases present in a new alloy. Although there is no database for Mg and Li containing HEAs due to insufficient amount of research published till date. There is need of First-principle DFT; MD simulation and Phase Field modelling to gain a better understanding of phases formation and understanding the factors that affects it.

Sanchez et al. used CALPHAD to predict phases present in $Al_{60}Cu_{10}Fe_{10}Cr_5Mn_5Ni_5Mg_5$ alloy and found the Cu, Mg, Ni doped Al_3Fe_4, Al_7Cu_4Ni, Al_8Mn_5, $D8_{10}$ Al-Cr alloy phase and $Mg_2Cu_6Al_5$ phases in the system [34]. The thermo-calc database successfully detected the phases present. Although, there are few discrepancies between the experimental results and Thermo-Calc calculations. CALPHAD predicted more constituent phases of which only few are reported by experiments. This suggests that despite having a slow cooling rate of solidification, total equilibrium state is not achieved in the manufacturing of the alloys. In general, this database has proven to be a good method for designing such HEAs [34].

8. Hydrogen storage behavior of Mg-containing HEAs

Amongst the growing need for energy and constant decline in non-renewable energy sources, renewable energy storage devices are gaining popularity. The demand for Hydrogen Storage Devices is also increasing for the same reason. Among many other alternatives of hydrogen storage principles, metal hydrides are considered as some ideal candidates as these do not require cryogenic cooling like liquid hydrogen storages and can absorb decent amount of hydrogen as hydrides due to the high bond strength between metal and hydrogen [54]. Most metal hydrides exhibit exothermic reactions during hydride formation, whereas; the desorption reaction is endothermic in most cases. So, external temperature rise is necessary to continue the storage cycle. The negative enthalpy and entropy values are responsible for the high bond strength between metal ions and hydrogen [40]. The ΔH=-74 kJ.$mol_{H_2}^{-1}$ and $\Delta S\sim$-135 J.$K^{-1}mol_{H_2}^{-1}$for MgH_2 [40, 54]. The major drawback of single metal hydrides is the low desorption of hydrogen leading to low cyclability [54]. Single metal hydrides exhibit a hydrogen to metal ratio, [H]/[M]= 0.6. To solve this issue, several approaches have been taken e.g. using two or more metals to form hydrides, using non-hydride compositions for hydrogen storage, using boro-hydrides and ultimately, MPEA, commonly known as HEA which are a matter of research now [54].

The hydrogen storage in HEAs is related to a reversible phase transformation upon hydrogen absorption. The complex crystal structure of HEAs can store hydrogen in both tetrahedral and octahedral voids simultaneously, resulting in a higher hydrogen storage capacity than any single or, binary metal hydride. TiVZrNbHf HEA has the hydrogen to metal ratio, [H]/[M]= 2.5 [57]. The hydrogen to metal ratio is 2 for single metal hydrides [41].

Zepon et. al. conducted a research on Mg containing HEA and found the hydrogen absorption capacity to be 1.2 wt% of the alloy. The HEA exhibits a BCC structure while it converts into an FCC structure upon hydrogen absorption [55]. The HEA was produced using high energy ball milling while the hydride was produced using reactive ball milling. The main reason for upgrading to Mg containing HEAs for hydrogen storage is because of the light weight of Mg which might reduce the weight of hydrogen storage devices in light duty fuel cell vehicles. Efficient hydrogen storage requires light weight storage devices for high gravimetric capacity [41].

Marcelo et. al. produced hydrides of MgVCr and MgVTiCrFe alloys using reactive milling and the MgVCr consisted of a BCC phase with presence of $\beta\text{-}MgH_2$ on 72 hours of reactive milling. While the MgVTiCrFe consisted of amorphous phase in this study. The hydrogen storage capacity was measured at 30°C, 150°C and 350°C temperature. The hydrogen storage capacity was observed to be extremely low at 30°C and 150°C while it increased at 350°C. The MgVTiCrFe alloy did not show very promising hydrogen storage characteristics but MgVCr is a promising candidate for this purpose exhibiting a reversible 0.95 wt% hydrogen absorption capacity [56].

Strozi et. al synthesized MgVAlCrNi HEA using high energy ball milling but the equiatomic compound showed a very low hydrogen storage capacity, so two non-equiatomic alloy, $Mg_{28}V_{28}Al_{19}Cr_{19}Ni_6$ and $Mg_{26}V_{31}Al_{31}Cr_6Ni_6$ were proposed and studied for hydrogen storage which also didn't show promising results. The non-equiatomic compositions were selected in such a way that the amount of Mg and V is increased, the solubility of hydrogen is increased and so does the lattice parameter because, the increase in lattice parameter indicates that there will be more available interstitial space for hydrogen absorption into the structure. This study prioritizes on the importance of the enthalpy of hydrogen solution on the hydrogen storage capacity of Mg containing HEAs [39]. The positive enthalpy of hydride formation for most of the elements present in these alloys is responsible for this low hydrogen storage behavior.

A very recent study by Montero et al. suggests the improvement in cycling behavior of TiVZrNb HEA upon introduction of Mg to it. After the 12th cycle, the absorption capacity reduces upto 2.41 wt% from the initial 2.8 wt%. The temperature where the maximum desorption occur, is 290°C for $Mg_{10}Ti_{10}V_{25}Zr_{10}Nb_{25}$, while it is 330°C for $Ti_{32.5}V_{27.5}Zr_{12.5}Nb_{27.5}$ [41]. As the desorption reaction is endothermic, it requires extermal temperature rise to promote the process. The lower the temperature when the desorption starts, the more economically viable and efficient the storage device will be. This particular composition shows the lowest temperature for the maximum desorption to occur. Observations from this study suggests that the $Mg_{10}Ti_{10}V_{25}Zr_{10}Nb_{25}$, alloy loses around 11 % of the hydrogen storage capacity in the second cycle, where the storage capacity reduces from 2.7% to

HEA composition	Hydrogen absorption capacity	Desorption temperature	[H]/[M]	Ref.
$MgZrTiFe_{0.5}Co_{0.5}Ni_{0.5}$	1.2 wt%	300°C	0.7	[55]
MgVCr	0.95 wt%	376°C		[56]
MgVTiCrFe	0.37 wt%	360°C		[56]
MgVAlCrNi	0.3 wt%	440°C	~0.15	[39]
$Mg_{10}Ti_{10}V_{25}Zr_{10}Nb_{25}$	2.8 wt%	290°C	1.7	[41]

Table 5.
Hydrogen absorption capacity of HEAs.

2.41%, whereas, $Ti_{32.5}V_{27.5}Zr_{12.5}Nb_{27.5}$ loses 28% of the hydrogen storage capacity during the first 4 cycles. This recent study suggests the enhancement in the reversable capacity of Mg containing HEAs than Mg-less HEAs. There are not many studies done on Mg containing HEAs for hydrogen storage applications and so the field is open for further research. But this particular study suggests a high potential of Mg containing MPEAs in this certain application.

The **Table 5** summarizes the findings.

9. Design, phase prediction and future of Mg HEA

This section gives an elementary guideline for design, phase prediction and future trends of magnesium containing HEAs. Before doing that it is more important to focus on possible applications of such alloys e.g. die casting, paneling of aircraft, weight reduction in automobiles and biomedical implants. Magnesium components are used in automobiles as instrument-panel beam, transfer case, steering components, air bag housing, seat tanks, fuel tank cover and radiator support. Typically Mg constitutes 4 kgs of normal cars weight which is fairly less. Mg can absorb 16 time more vibrations compared to Al, hence alloys of Mg can be used for shock absorbing applications. The major reason for limited use of Magnesium alloys is low strength compared to Aluminum alloys and Steels. The concept of maximization of entropy via mixing multiple elements in near equiatomic ratios to creating "base" or solvent less alloy. Till date, various HEAs have proven their strength. The implementation of this system to Mg could be breakthrough for automotive, space, missiles and aircraft industry. To do so authors provide a preliminary scientific approach to develop Mg HEAs.

Mg has high solubility with Li ($\sim$ 17 at. %), Al ($\sim$12 at. %), In ($\sim$19 at. %) and it is completely soluble with Cadmium. Although Li and In are soft metals like Mg. It has very high tendency to form intermetallic compounds with metals such as Al, Zn, Cu, Y, Zr and Ca etc. It forms the famous quasicrystals when alloyed with Mn. Mg finds place in a wide spectrum of alloys, compounds and systems. Perhaps it is most interesting element, yet to be studied thoroughly in the complex concentrated systems or HEAs. The short range order in HEAs is recently reported to be a core effect for strengthening in such alloys. Mg provides a great avenue for tailoring heterogeneities in HEAs. In the light of limited data of Mg containing HEAs with only 35 compositions, it is hard for authors to suggest the role of thermodynamic and kinetic criteria to develop Mg-HEAs. There is an insufficient data to develop an analogy on the phase development and phase evolution of Mg containing HEAs.

It has been understood that critical values of theoretical parameters used for conventional heavy HEAs are not sufficient and effective in case of LHEAs containing especially Mg owing to its larger size, high $|\Delta Hmix|$ with most of elements leading to immiscibility ($\Delta Hmix$ is +ve) and intermetallic formation ($\Delta Hmix$ is –ve). Yang et. al. proposed new limits to the critical value for light weight HEAs. The modified threshold values suggest that SS will form at $\Delta Hmix \in [-1, 5]$ kJ/mol; $\delta < 4.5\%$ and $\Omega > 10$. Mg amongst all the elements in the reported alloys has higher radius due to which poly-disparity constant "δ" value is higher and hence does not support complete SS formation. It is evident that $\Omega > 10$ is an ideal condition for SS formation but in various cases a pure SS is obtained when $\Omega < 10$, which is an unknown at present and shall be governed by the effect of individual elements for instance; in few of such alloys Mg was alloyed with refractory elements such as Mo and Nb and in this case, δ had a high value. In few alloys the effect of high configurational entropy of mixing is overshadowed by high negative enthalpy of mixing and high atomic mismatch. HEAs containing Mg, Li and Al open a new avenue for scientific research and new outlook for understanding of such complex systems. It is simply understood that if an alloy contain more elements with HCP crystal structure (Mg, Ti, Sc, Co, Zn, Cd, Zr and Y), it should probably result in HCP structure of alloy. This is evident from the authors analysis. It can be concluded that for designing HEAs with Mg, Li and Al, ΔHmix should be given priority over entropy of mixing. It is crucial to study the possible binary and ternary combinations out of the sought HEA composition.

For the design of LHEAs, Mg alone should not be essentially alloyed with Al and/or Li. At the same time elements soluble with Al or Li can be used to further increase the disordered structure. As it has been also observed that the presence of d orbital element is necessary to obtain high entropy effect in the alloy [35]. To understand the phases, pseudo binary phase diagrams can also be produced using Thermo-Calc, molecular dynamics simulations or extrapolating the experimental data. Due to lack of experimental data, the extrapolation may not be accurate. Phase formation and phase stabilization can be understood by studying the thermodynamic parameters. Few graphs have been plotted shown in **Figures 8–10** using the basic thermodynamic parameters such as $\boldsymbol{\Delta H_{mix}}, \boldsymbol{\Delta S_{mix}}, \boldsymbol{\delta}, \boldsymbol{\chi}, \boldsymbol{\Omega}$ and VEC from the data shown in **Table 4**, which are considered as the phase formation parameters. Values of these parameters can be calculated from the equations given in the **Table 6** below.

Table 7 shows the type of compound formation based on the values of $\boldsymbol{\Delta H_{mix}}$ *and* $\boldsymbol{\Delta S_{mix}}$. Larger negative values of mixing enthalpy may lead to the compound formation. For solid solution formation, higher negative values of entropy is required. Positive values of mixing enthalpy and mixing entropy may lead to the elemental or compound segregation within alloys.

Figure 8 shows the graph between the $\boldsymbol{\Delta H_{mix}}$ and $\frac{\Delta S_{mix}}{R}$, where R is the Gas constant of the data shown in **Table** 5. Graph shows that pure solid solution is stable at entropy higher than 1.13 kJ/mol. IM are observed in both low entropies and higher entropy region as well. Pure solid solution is observed at medium negative values of $\boldsymbol{\Delta H_{mix}}$.One exceptional is when solid solution has been obtained at higher

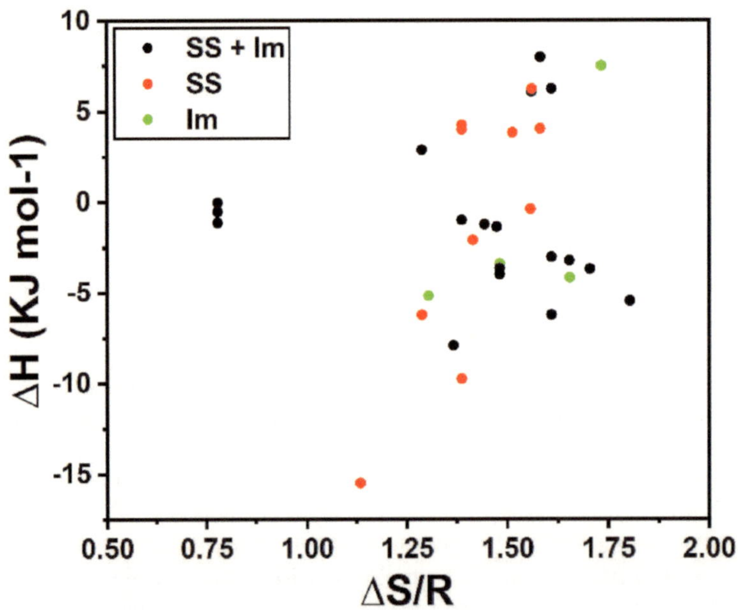

Figure 8.
*Graph between enthalpy and entropy from the data shown in **Table 4**.*

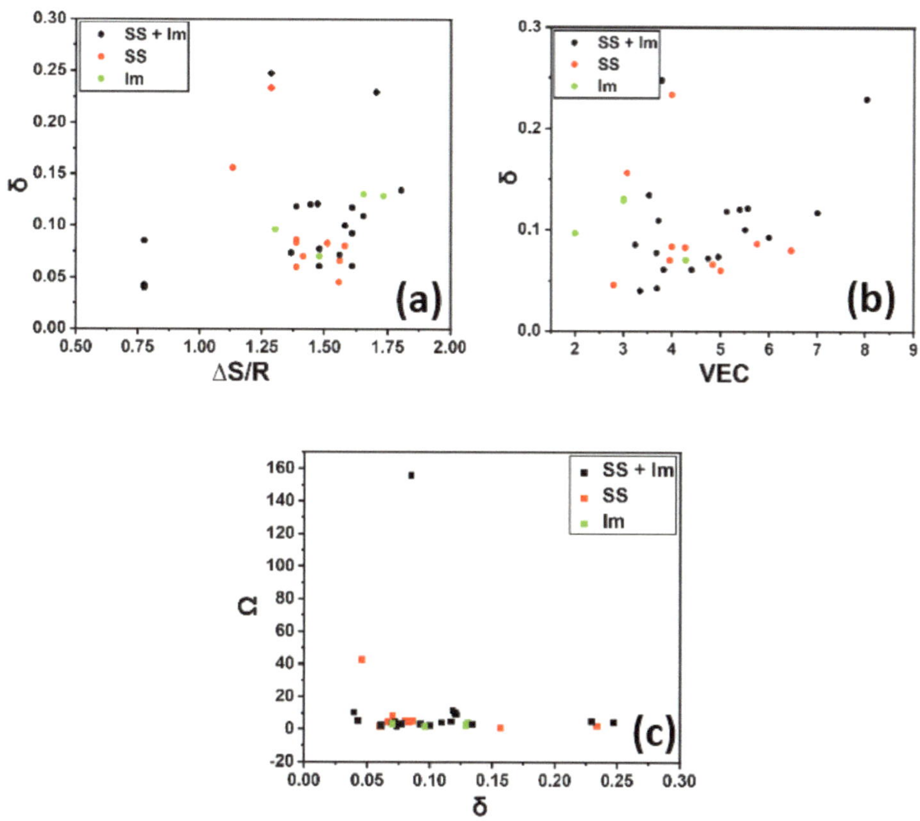

Figure 9.
(a) Graph between atomic mismatch factor and VEC; (b) graph between the atomic mismatch factor and omega; (c) graph between atomic mismatch factor and VEC. Data has been taken from **Table** *5.*

negative value of ΔH_{mix}. That is due to the presence of Ca in the alloy. Mg tends to form stable compounds with the Ca: Presence of Ca enhances the solid solution formation in the presence of Mg.

Effect of δ is need to be understood properly. As per the Hume Rothery solid solution guidelines, the atomic mismatch factor must be minimum ($\delta \leq 15\%$). The atomic size of light weight elements, Al, Mg, In, Li, Ca is higher as compared to the d block and other elements, due to which there is higher tendency of phase separation. **Figure 9a-c** shows effect of δ in solid solution formation. The data is used from **Table 5** to plot the graphs against $\frac{\Delta S_{mix}}{R}$ and VEC. **Figure 9a** shows that the pure solid solution are observed only between $0.043 \leq \delta \leq 0.24$. Pure intermetallic compounds are observed between $0.067 \leq \delta \leq 0.133$, and combination of intermetallic and solid solution is observed for rest of the regions as solid solutions are stable only at higher values of entropy, and medium atomic mismatch factor.

Figure 9b shows the effect of δ with the Ω; Ideally, for solid solution formation the value of Ω should be near 1.1 [60], but solid solutions are also observed at higher values. **Figure 9b** shows that solid solution forms within $0.41 \leq \delta \leq 0.24$ and intermetallic compounds are observed at medium values of $0.066 \leq \delta \leq 0.133$. Combination of solid solution and intermetallic compounds are observed between $0.037 \leq \delta \leq 0.25$. Similarly **Figure 9c** shows that solid solutions are stable between $0.043 \leq \delta \leq 0.24$, longer range of atomic mismatch factor, and intermetallic compounds are observed between $0.067 \leq \delta \leq 0.13$. Combination of solid solution and intermetallic compound is observed at all other regions. These graphs show that

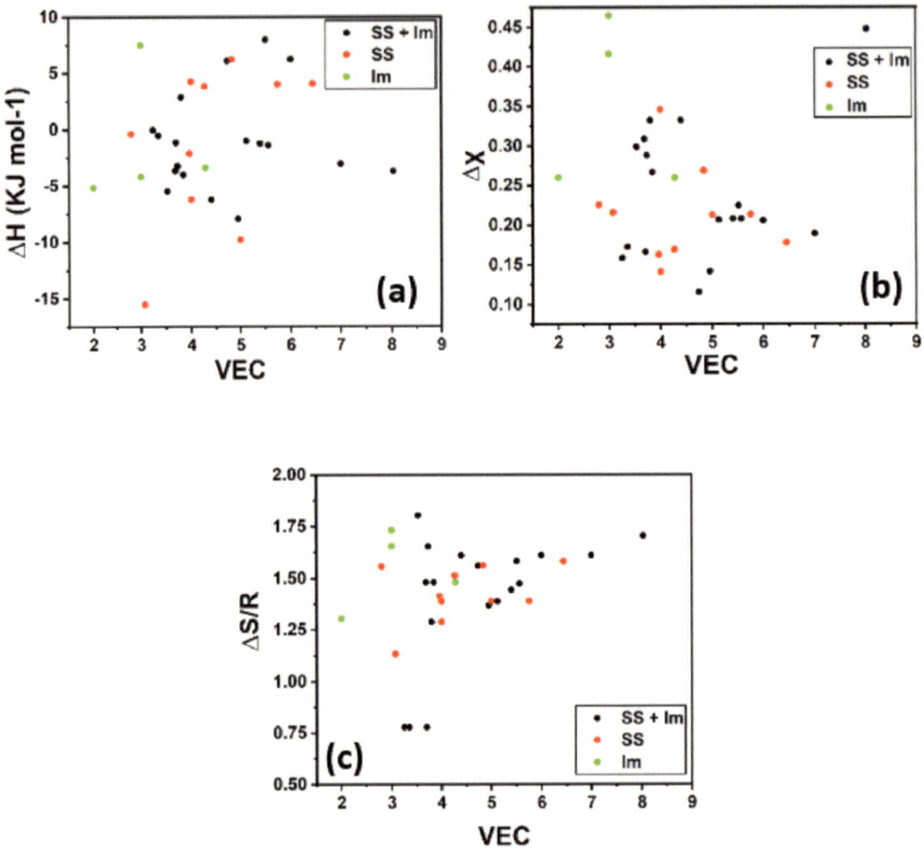

Figure 10.
(a) Graph between VEC and enthalpy; (b) graph between VEC and electro-negativity; (c) graph between VEC and entropy. Data has been taken from ***Table 5****.*

1.	Atomic size 2difference	$\delta = \sqrt{\left(\Sigma_i^n c_i \left(1 - \frac{r_i}{\bar{r}}\right)^2\right)}$ C_i is atomic percentage	[35]
2.	Enthalpy	$\Delta H_{mix} = \Sigma_{i=1, i \neq j}^n \Omega_{ij} C_i C_j$ C_i and C_j is atomic percentage	[35]
3.	Macro states: Statistical Thermodynamic parameter	$\Omega = \frac{T_m \Delta S_{mix}}{\lvert \Delta H_{mix} \rvert}$ T_m is the melting point of n elements.	[35]
4.	Melting point of n elements	$T_m = \Sigma_{i=1} n_{C_i} (T_m)_i$ n is the number of elements. C_I is the atomic percentage; T_{mi} is the melting point of ith element.	[35]
5.	Pauling electronegativity difference	$\Delta\chi = \sqrt{\Sigma_{i=1}^n c_i (\chi_i - \bar{\chi})^2}$ C_i is the atomic percentage of ith element. $\bar{\chi}$ is average electronegativity χ is the electronegativity of ith element	[35]
6.	VEC	VEC = $\Sigma_{i=1}^n c_i (VEC)_i$ C_i is the atomic percentage of ith element.	[35]

Table 6.
List of equations for calculation of thermodynamic parameters.

Mixing Enthalpy	Mixing Entropy	Gibbs Free Energy	Types of phase
$\Delta H_{mix} = 0$	$\Delta S_{mix} = 0$	$\Delta G_{mix} = 0$	Elemental Phases
$\Delta H_{mix} \ll 0$ (*Large negative*)	$\Delta S_{mix} = 0$	$\Delta G_{mix} \ll 0$ (*Large negative*)	Compounds
$\Delta H_{mix} < 0$	$\Delta S_{mix} < 0$ (*medium negative*)	$\Delta G_{mix} \ll 0$ (*Large negative*)	Intermediate phase
$\Delta H_{mix} < 0$ (*medium negative*)	$\Delta S_{mix} < -1.5R$	$\Delta G_{mix} \ll 0$ (*Large negative*)	Random solid solution

Table 7.
Types of phase based on the values of thermodynamic parameters, Gibbs free energy ΔG_mix, mixing enthalpy ΔH_mix and mixing Entropy ΔS_mix [60].

there is high possibility of ordering within a high entropy solid solution, as also said above [35]. Simple solid solution of HEA could be obtained between $0.041 \leq \delta \leq 0.24$. Solid solution formation is also depending on the values of Valence Electron Concentration (VEC), which are being discussed in section 10.

Figure 10 shows the effect of VEC on solid solution formation in LHEAs. VEC is the total number of free electrons including the d-orbital electrons that can participate in the formation of chemical bond. Generally, VEC also helps to understand the type of phase formation in the alloy. FCC phases are found to be stable at VEC 8, and BCC phases are stable at VEC<6.87. Combination of BCC + FCC are stable for the value of VEC between 6.87 < VEC < 8.

Figures 9b and **10a-c**, shows the effect of VEC on solid solution formation with respect to $\boldsymbol{\Delta H_{mix}}$, χ and $\frac{\Delta S_{mix}}{R}$. All of the three graphs shows exact same values for solid solution, Intermetallic compound and combination of solid solution and intermetallic compound formation. Solid solution is observed between 2.7< VEC<6.5; Intermetallic compounds are observed between 1.92<VEC<4.35 and combination of solid solution and intermetallic compounds are observed between 2.8<VEC<8.1.

All the parameters can be understood as the pure ss of LHEA will form for the following parameters.

1. $-15 < \Delta H_{mix} < 6.5$
2. $1.1 < \frac{\Delta S_{mix}}{R} < 1.6$
3. $0.041 \leq \delta \leq 0.24$
4. $2.7 < \text{VEC} < 6.5$

If an alloy follows these certain condition, we can obtain the single phase LHEAs.

It is observed that alloys with Ca have a higher tendency to form IM, similar result have been observed by Nagase et. al. The melting point (MP) of constituting elements also plays a significant role, as a higher difference in MP leads to segregation upon cooling. It should also be accepted that formation of IM cannot avoided in Complex concentrated systems as there is always a possibility of two random elements having higher negative enthalpy of mixing. It is important to note that till date no Mg containing HEA has shown LPSO. LPSO containing Mg alloys if possible could have exhibited better properties and stability. VEC rule is well-established in HEAs, but its drawback is that it does not discuss about lattice other than BCC and FCC, in the case of Mg HEAs, most of the lower density alloys crystallizes in HCP.

10. Applications

The unique compositions of HEAs give rise to a new set of properties which makes every high entropy alloy unique. Mg containing high entropy alloys have shown promising features which make them unique for the following applications.

1. Aerospace alloys and alloys in motor vehicles as a replacement of Al based alloys: Aluminium is already a popular materials for construction of motor vehicles and even aerospace materials. Mg being lighter than Al in fact acts as a means of reducing the density of the alloy even more.

2. High strength and corrosion resistant applications.

3. Hydrogen storage devices: The details of the hydrogen storage behaviour of Mg containing HEAs are already discussed in section 6. According to **Table 6**, $MgZrTiFe_{0.5}Co_{0.5}Ni_{0.5}$ and $Mg_{10}Ti_{10}V_{25}Zr_{10}Nb_{25}$ are the most promising candidates for this purpose.

Conflict of interest

The authors declare no conflict of interest.

Author details

Prince Sharma[1*], Nushrat Naushin[2], Sahil Rohila[3] and Abhishek Tiwari[1]

1 Department of Metallurgical and Materials Engineering, Indian Institute of Technology, Kharagpur, WB, India

2 Department of Materials Science and Engineering, Khulna University of Engineering and Technology, Khulna, Bangladesh

3 Department of Materials Science and Metallurgical Engineering, Indian Institute of Technology, Sangareddy, Telangana, India

*Address all correspondence to: princesharma@alumni.iith.ac.in

References

[1] G.R. Josyer, Maharshi Bharadwaaj's "The Vymanika Shastra," English Tr, Internaitonal Academy of Sanskrit Research, Mysore-4. eastward ones opened., Mysore, 1973.

[2] B. Ravi, Investment Casting Development: Ancient and Modern Approaches, n.d.

[3] B. Cantor, I.T.H. Chang, P. Knight, A.J.B. Vincent, Microstructural development in equiatomic multicomponent alloys, Mater. Sci. Eng. A. 375–377 (2004) 213–218. https://doi.org/10.1016/j.msea.2003.10.257.

[4] J.-W. Yeh, S.-K. Chen, S.-J. Lin, J.-Y. Gan, T.-S. Chin, T.-T. Shun, C.-H. Tsau, S.-Y. Chang, Nanostructured High-Entropy Alloys with Multiple Principal Elements: Novel Alloy Design Concepts and Outcomes, Adv. Eng. Mater. 6 (2004) 299–303. https://doi.org/10.1002/adem.200300567.

[5] D.B. Miracle, O.N. Senkov, A critical review of high entropy alloys and related concepts, Acta Mater. 122 (2017) 448–511.

[6] P. Sharma, Electro-spark Coating of High Entropy Alloy on Steel and Inconel Superalloy for Surface Hardening and Oxidation Protection, Master Thesis, Indian Institute of Technology Hyderabad 2020. https://doi.org/10.13140/RG.2.2.13942.32327/1.

[7] X. Yang, Y. Zhang, Prediction of high-entropy stabilized solid-solution in multi-component alloys, Mater. Chem. Phys. 132 (2012) 233–238. https://doi.org/10.1016/j.matchemphys.2011.11.021.

[8] J. Dąbrowa, M. Zajusz, W. Kucza, G. Cieślak, K. Berent, T. Czeppe, T. Kulik, M. Danielewski, Demystifying the sluggish diffusion effect in high entropy alloys, J. Alloys Compd. 783 (2019) 193–207.

[9] S. V Divinski, A. V Pokoev, N. Esakkiraja, A. Paul, A mystery of" sluggish diffusion" in high-entropy alloys: the truth or a myth?, in: Diffus. Found., Trans Tech Publ, 2018: pp. 69–104.

[10] K. Kulkarni, G.P.S. Chauhan, Investigations of quaternary interdiffusion in a constituent system of high entropy alloys, AIP Adv. 5 (2015) 97162.

[11] K.-Y. Tsai, M.-H. Tsai, J.-W. Yeh, Sluggish diffusion in Co–Cr–Fe–Mn–Ni high-entropy alloys, Acta Mater. 61 (2013) 4887–4897.

[12] K. Jin, C. Zhang, F. Zhang, H. Bei, Influence of compositional complexity on interdiffusion in Ni-containing concentrated solid-solution alloys, Mater. Res. Lett. 6 (2018) 293–299.

[13] M. Vaidya, S. Trubel, B.S. Murty, G. Wilde, S. V Divinski, Ni tracer diffusion in CoCrFeNi and CoCrFeMnNi high entropy alloys, J. Alloys Compd. 688 (2016) 994–1001.

[14] M. Vaidya, K.G. Pradeep, B.S. Murty, G. Wilde, S. V Divinski, Radioactive isotopes reveal a non sluggish kinetics of grain boundary diffusion in high entropy alloys, Sci. Rep. 7 (2017) 1–11.

[15] M. Vaidya, K.G. Pradeep, B.S. Murty, G. Wilde, S. V Divinski, Bulk tracer diffusion in CoCrFeNi and CoCrFeMnNi high entropy alloys, Acta Mater. 146 (2018) 211–224.

[16] D. Gaertner, J. Kottke, G. Wilde, S. V Divinski, Y. Chumlyakov, Tracer diffusion in single crystalline CoCrFeNi and CoCrFeMnNi high entropy alloys, J. Mater. Res. 33 (2018) 3184–3191.

[17] J. Zhang, G.M. Muralikrishna, A. Asabre, Y. Kalchev, B. Butz, J. Müller, S.

Hilke, H. Rösner, G. Laplanche, S. V Divinski, Tracer Diffusion in the Σ-Phase of the CrMnFeCoNi System, Available SSRN 3688786. (n.d.).

[18] J. Dąbrowa, W. Kucza, G. Cieślak, T. Kulik, M. Danielewski, J.W. Yeh, Interdiffusion in the FCC-structured Al-Co-Cr-Fe-Ni high entropy alloys: Experimental studies and numerical simulations, J. Alloys Compd. 674 (2016) 455–462. https://doi.org/10.1016/j.jallcom.2016.03.046.

[19] V.M. Nadutov, V.F. Mazanko, S.Y. Makarenko, Tracer Diffusion of Cobalt in High-Entropy Alloys AlxFeNiCoCuCr, Металлофизика и Новейшие Технологии. (2017).

[20] S. Ranganathan, Alloyed pleasures: Multimetallic cocktails, Curr. Sci. 85 (2003) 1404–1406.

[21] R.S. Mishra, R.S. Haridas, P. Agrawal, High entropy alloys – Tunability of deformation mechanisms through integration of compositional and microstructural domains, Mater. Sci. Eng. A. 812 (2021) 141085. https://doi.org/https://doi.org/10.1016/j.msea.2021.141085.

[22] F.X. Zhang, S. Zhao, K. Jin, H. Xue, G. Velisa, H. Bei, R. Huang, J.Y.P. Ko, D. C. Pagan, J.C. Neuefeind, W.J. Weber, Y. Zhang, Local Structure and Short-Range Order in a NiCoCr Solid Solution Alloy, Phys. Rev. Lett. 118 (2017) 205501. https://doi.org/10.1103/PhysRevLett.118.205501.

[23] R. Zhang, S. Zhao, J. Ding, Y. Chong, T. Jia, C. Ophus, M. Asta, R.O. Ritchie, A.M. Minor, Short-range order and its impact on the CrCoNi medium-entropy alloy, Nature. 581 (2020) 283–287. https://doi.org/10.1038/s41586-020-2275-z.

[24] M.M. Nygård, W.A. Sławiński, G. Ek, M.H. Sørby, M. Sahlberg, D.A. Keen, B.C. Hauback, Local order in high-entropy alloys and associated deuterides–a total scattering and Reverse Monte Carlo study, Acta Mater. 199 (2020) 504–513.

[25] O. Maulik, D. Kumar, S. Kumar, S.K. Dewangan, V. Kumar, Structure and properties of lightweight high entropy alloys: A brief review, Mater. Res. Express. 5 (2018) 52001. https://doi.org/10.1088/2053-1591/aabbca.

[26] E. Abe, Fascinating LPSO-Structured Mg Alloys, in: Magnes. Technol. 2016, Springer International Publishing, 2016: pp. 11–12. https://doi.org/10.1007/978-3-319-48114-2_4.

[27] L.W. Cheah, Cars on a Diet: The Material and Energy Impacts of Passenger Vehicle Weight Reduction in the U.S, 2001.

[28] M. Gupta, N.M.L. Sharon, Magnesium, Magnesium Alloys, and Magnesium Composites, Wiley, 2011. https://doi.org/10.1002/9780470905098.

[29] M. Mezbahul-Islam, A. Mostafa, M. Medraj, Essential magnesium alloys binary phase diagrams and their thermochemical data, J. Mater. 2014 (2014).

[30] K.K. Sankaran, R.S. Mishra, Chapter 7 - Magnesium Alloys, in: K.K. Sankaran, R.S. Mishra (Eds.), Metall. Des. Alloy. with Hierarchical Microstruct., Elsevier, 2017: pp. 345–383. https://doi.org/10.1016/B978-0-12-812068-2.00007-2.

[31] R. Li, J.C. Gao, K. Fan, Microstructure and mechanical properties of MgMnAlZnCu high entropy alloy cooling in three conditions, in: Mater. Sci. Forum, Trans Tech Publ, 2011: pp. 235–241.

[32] R. Li, J.C. Gao, K. Fan, Study to microstructure and mechanical properties of Mg containing high entropy alloys, in: Mater. Sci. Forum, Trans Tech Publ, 2010: pp. 265–271.

[33] K.S. Tun, M. Gupta, Microstructural evolution in MgAlLiZnCaY and MgAlLiZnCaCu multicomponent high entropy alloys, in: Mater. Sci. Forum, Trans Tech Publications Ltd, 2018: pp. 183–187. https://doi.org/10.4028/www.scientific.net/MSF.928.183.

[34] J.M. Sanchez, I. Vicario, J. Albizuri, T. Guraya, J.C. Garcia, Phase prediction, microstructure and high hardness of novel light-weight high entropy alloys, J. Mater. Res. Technol. 8 (2019) 795–803.

[35] X. Yang, S.Y. Chen, J.D. Cotton, Y. Zhang, Phase Stability of Low-Density, Multiprincipal Component Alloys Containing Aluminum, Magnesium, and Lithium, Jom. 66 (2014) 2009–2020. https://doi.org/10.1007/s11837-014-1059-z.

[36] Z. Gu, P. Mao, Y. Gou, Y. Chao, S. Xi, Microstructure and properties of MgMoNbFeTi2Yx high entropy alloy coatings by laser cladding, Surf. Coatings Technol. 402 (2020) 126303. https://doi.org/10.1016/j.surfcoat.2020.126303.

[37] K.M. Youssef, A.J. Zaddach, C. Niu, D.L. Irving, C.C. Koch, A novel low-density, high-hardness, high-entropy alloy with close-packed single-phase nanocrystalline structures, Mater. Res. Lett. 3 (2014) 95–99. https://doi.org/10.1080/21663831.2014.985855.

[38] O. Maulik, N. Patra, D. Bhattacharyya, S.N. Jha, V. Kumar, Local atomic structure investigation of AlFeCuCrMgx (0.5, 1, 1.7) high entropy alloys: X-ray absorption spectroscopy study, Solid State Commun. 252 (2017) 73–77. https://doi.org/10.1016/j.ssc.2017.01.018.

[39] R.B. Strozi, D.R. Leiva, J. Huot, W.J. Botta, G. Zepon, Synthesis and hydrogen storage behavior of Mg–V–Al–Cr–Ni high entropy alloys, Int. J. Hydrogen Energy. 46 (2021) 2351–2361.

[40] I.P. Jain, C. Lal, A. Jain, Hydrogen storage in Mg: A most promising material, Int. J. Hydrogen Energy. 35 (2010) 5133–5144. https://doi.org/10.1016/j.ijhydene.2009.08.088.

[41] J. Montero, G. Ek, M. Sahlberg, C. Zlotea, Improving the hydrogen cycling properties by Mg addition in Ti-V-Zr-Nb refractory high entropy alloy, Scr. Mater. 194 (2021) 113699.

[42] P. Chauhan, S. Yebaji, V.N. Nadakuduru, T. Shanmugasundaram, Development of a novel light weight Al35Cr14Mg6Ti35V10 high entropy alloy using mechanical alloying and spark plasma sintering, J. Alloys Compd. 820 (2020) 153367.

[43] O. Maulik, V. Kumar, Synthesis of AlFeCuCrMgx (x = 0, 0.5, 1, 1.7) alloy powders by mechanical alloying, Mater. Charact. 110 (2015) 116–125. https://doi.org/10.1016/j.matchar.2015.10.025.

[44] N. Singh, Y. Shadangi, N.K. Mukhopadhyay, Phase Evolution and Thermal Stability of Low-Density MgAlSiCrFe High-Entropy Alloy Processed Through Mechanical Alloying, Trans. Indian Inst. Met. 73 (2020) 2377–2386. https://doi.org/10.1007/s12666-020-02039-y.

[45] Y. Zhang, Y.J. Zhou, J.P. Lin, G.L. Chen, P.K. Liaw, Solid-Solution Phase Formation Rules for Multi-component Alloys, Adv. Eng. Mater. 10 (2008) 534–538. https://doi.org/10.1002/adem.200700240.

[46] A. Takeuchi, A. Inoue, Calculations of Mixing Enthalpy and Mismatch Entropy for Ternary Amorphous Alloys, Mater. Trans. JIM. 41 (2000) 1372–1378. https://doi.org/10.2320/matertrans1989.41.1372.

[47] A. Takeuchi, A. Inoue, Classification of Bulk Metallic Glasses by Atomic Size Difference, Heat of Mixing and Period of Constituent Elements and Its

Application to Characterization of the Main Alloying Element, Mater. Trans. 46 (2005) 2817–2829. https://doi.org/10.2320/matertrans.46.2817.

[48] A.R. Miedema, On the heat of formation of solid alloys. II, J. Less-Common Met. 46 (1976) 67–83. https://doi.org/10.1016/0022-5088(76)90180-6.

[49] R. Boom, F.R. De Boer, A.R. Miedema, On the heat of mixing of liquid alloyS-II, J. Less-Common Met. 46 (1976) 271–284. https://doi.org/10.1016/0022-5088(76)90215-0.

[50] A.R. Miedema, F.R. de Boer, R. Boom, Model predictions for the enthalpy of formation of transition metal alloys, Calphad. 1 (1977) 341–359. https://doi.org/10.1016/0364-5916(77)90011-6.

[51] A.R. Miedema, R. Boom, F.R. De Boer, On the heat of formation of solid alloys, J. Less-Common Met. 41 (1975) 283–298. https://doi.org/10.1016/0022-5088(75)90034-X.

[52] A.R. Miedema, F.R. De Boer, R. Boom, Predicting heat effects in alloys, Phys. B+C. 103 (1981) 67–81. https://doi.org/10.1016/0378-4363(81)91003-2.

[53] K.S. Tun, A. Kumar, M. Gupta, Introducing a High Performance Mg-Based Multicomponent Alloy as an Alternative to Al-Alloys, Front. Mater. 6 (2019). https://doi.org/10.3389/fmats.2019.00215.

[54] L.J. Bannenberg, M. Heere, H. Benzidi, J. Montero, E.M. Dematteis, S. Suwarno, T. Jaroń, M. Winny, P.A. Orłowski, W. Wegner, A. Starobrat, K.J. Fijałkowski, W. Grochala, Z. Qian, J.P. Bonnet, I. Nuta, W. Lohstroh, C. Zlotea, O. Mounkachi, F. Cuevas, C. Chatillon, M. Latroche, M. Fichtner, M. Baricco, B.C. Hauback, A. El Kharbachi, Metal (boro-) hydrides for high energy density storage and relevant emerging technologies, Int. J. Hydrogen Energy. 45 (2020) 33687–33730. https://doi.org/10.1016/j.ijhydene.2020.08.119.

[55] G. Zepon, D.R. Leiva, R.B. Strozi, A. Bedoch, S.J.A. Figueroa, T.T. Ishikawa, W.J. Botta, Hydrogen-induced phase transition of MgZrTiFe0. 5Co0. 5Ni0. 5 high entropy alloy, Int. J. Hydrogen Energy. 43 (2018) 1702–1708.

[56] M.O. de Marco, Y. Li, H.-W. Li, K. Edalati, R. Floriano, Mechanical Synthesis and Hydrogen Storage Characterization of MgVCr and MgVTiCrFe High-Entropy Alloy, Adv. Eng. Mater. 22 (2020) 1901079.

[57] M.M. Nygård, G. Ek, D. Karlsson, M. Sahlberg, M.H. Sørby, B.C. Hauback, Hydrogen storage in high-entropy alloys with varying degree of local lattice strain, Int. J. Hydrogen Energy. 44 (2019) 29140–29149.

[58] T. Nagase, A. Terayama, T. Nagaoka, N. Fuyama, T. Sakamoto, Alloy Design and Fabrication of Ingots of Al–Mg–Li–Ca Light-Weight Medium Entropy Alloys, Mater. Trans. 61 (2020) 1369–1380. https://doi.org/10.2320/matertrans.F-M2020825.

[59] B. Cantor, Multicomponent and High Entropy Alloys, Entropy. 16 (2014) 4749–4768.

[60] M.C. Gao, P.K. Liaw, J.W. Yeh, Y. Zhang, High-entropy alloys: Fundamentals and applications, 2016. https://doi.org/10.1007/978-3-319-27013-5.

Chapter 3

Metal Forming of Magnesium Alloys for Various Applications

Romana Ewa Śliwa

Abstract

The chapter presents an analysis of selected magnesium alloys as structural materials to be used in production of parts as well as their technological parameters in some manufacturing processes: metal forming and joining. Taking into account the analysis of microstructure and mechanical properties of conventional and new magnesium alloys and requirements of their possible applications (aviation, automotive, sport, etc.), the study of forming parts/products based on description of plastic formability of magnesium alloys in the processes of bulk metal forming (forging, extrusion, KOBO extrusion, rolling) and joining (friction stir welding) has been presented. Upsetting test, backward extrusion, and KOBO extrusion of complex cross-sectional profiles and forging process were conducted using magnesium alloys such as AZ31, AZ61, AZ80, WE 43, and Mg alloy with Li for the production of thin-walled profiles and forged parts. The range of temperatures and extrusion rate for manufacturing of these profiles were determined. Tests also covered the analysis of microstructure of Mg alloys in the initial state as well as after the extrusion process. The recommendations for realization of metal forming and joining processes of selected magnesium alloys have been presented.

Keywords: magnesium alloys, plastic formability, forging, extrusion, KOBO extrusion, rolling, joining, friction stir welding

1. Introduction

Mg-alloys are being considered as one of the most versatile material choices among the structural materials that exhibit both energy efficiency and environmental benefits. Mg-based materials (alloys and composites) have enormous and unlimited potential to replace aluminum, steel, and structural plastics in diverse industrial and commercial sectors such as automotive, aviation, defense, biomedical, sporting equipment, consumer electronics, etc. These applications result from the need to use magnesium and its alloys as a material with favorable physical properties, especially high relative strength (Rm/ρ). Due to the good casting properties of magnesium, it was used primarily in cast structural elements. Wrought alloys have been used on a smaller scale so far, but material research and plastic formability processes to produce semi-finished products from magnesium alloys are currently under intensive development. The development is caused by the possibility of using various types of light structures in the construction, including vehicles made of magnesium-based materials, for which, for example, thin sheets are the

IntechOpen

basic initial material. This is mainly due to the attempt to reduce the weight of the structure and ensure adequate strength.

The interest in magnesium alloys for various structural elements, e.g., for the aviation industry, dates back to the 1940s. Examples of applications concerned, for example, the Northrop XP-56 plane, in which virtually all parts exposed to elevated temperature were made of magnesium and its alloys [1], B-36 bomber or [2], the S55 helicopter by Westland Aircraft Ltd. (1950 r.) [1, 3].

In later years, the use of magnesium alloys for aircraft structural elements was significantly limited, which was mainly due to the rapid corrosion of alloys (the main disadvantage), opinions about the flammability of magnesium, low metallurgical purity, low strength, high processing costs through plastic formability and poor machinability [4–7].

Since the beginning of the twenty-first century, there has been a renewed, significant increase in interest in magnesium alloys for applications in the aerospace industry. It results from the development of new coatings that can protect alloys against corrosion, new alloys, new technologies of obtaining semi-finished products by casting methods, and the improvement of various plastic forming technologies, which significantly improves the properties of the product and allows the use of these alloys wherever engineers look for very light construction elements, relatively durable and with anticorrosion protection at the same time [8–10].

The main problem in the development of techniques for processing magnesium alloys with metal forming methods is low formability [4, 11]. Low deformability at room temperature as well as temperature increased to 200°C of magnesium alloys result from a limited number of hexagonal lattice slip systems. The test results show that the microstructure of Al-Mg-Zn alloys deformed at temperatures up to 200°C shows bands slip and deformation twins [11]. In the range from 200 to 300°C, limited dynamic recovery and the formation of nuclei of dynamic recrystallization are observed. Continuous dynamic recrystallization takes place above 300°C, which results in an almost twofold increase in formability.

Metal forming of magnesium and its alloys is carried out, depending on the content of alloying elements, only in a narrow temperature range. Grains with an average diameter of up to 10 μm were obtained in magnesium alloys by thermoplastic treatment. The fragmentation of the grains below 10 μm is only achieved by introducing large deformations. The use of unconventional methods of deformation allows for obtaining the grinding of magnesium alloys to sub-micrometric or nanometric sizes. Therefore, these deformation methods constitute a very significant support for conventional forming methods [12–14].

In magnesium alloys, deformation processes are carried out at an elevated temperature. Therefore, it is practically impossible to obtain nanometric grain sizes obtained by the development of shear bands. The most common methods of large deformation, leading to the grinding of the grains of magnesium alloys, are equal channel angular pressing (ECAP) [15, 16] or hydrostatic extrusion [14, 17].

Basic magnesium alloys for metal forming contain up to 8% Al and the addition of Mn (up to 2%), Zn (usually up to 1.5%), Si (about 0.1%), and traces of Cu, Ni, Fe. There are three basic groups of magnesium alloys for metal forming. The first group includes mainly alloys with the addition of aluminum, zinc, and manganese. The second group includes alloys containing mainly the elements Zn, RE, Y, Zr, Th, and the third group, which is in the phase of intensive research, consists of new ultralight alloys containing lithium.

Magnesium alloys for metal forming are still used to a relatively small extent, which results from technological difficulties in metal forming and high production costs [13, 18, 19]. The main disadvantage of magnesium and its alloys is low deformability at ambient temperature, which results from the type of crystal lattice.

At ambient temperature, only one in-plane glide system (0001) is active. In addition to low temperature slip, a significant amount of twinning is observed in magnesium alloys.

Hot forming of magnesium alloys, depending on the chemical composition and deformability, is performed using the following methods [20, 21]:

- rolling (mainly grades from the group of Mg Al-Zn and Mg-Zn-Mn alloys, as well as new alloys of the Mg-Th- (Mn or Zr) and Mg-Li-Al alloys,
- open die and die forging,
- extrusion (alloys AZ31 (Mg-Al-Zn), AZ61 (Mg-Al-Zn), ZM21 (Mg-Zn-Mn),
- KOBO extrusion (AZ31, AZ 61 Az80, WE43)
- sheet metal stamping after the rolling process in heated dies.

The formability effect of magnesium alloys has been recently also used in a relatively new solid-state joining process, under the influence of frictional heat, i.e., Friction Stir Welding (FSW). There is no melting involved in the process unlike conventional Fusion welding processes. This method gives very good results in the creation of strong joints, competitive to other joining techniques that require additional joining materials (e.g., riveting, bolted connections, welding, and others). Joining elements (e.g., sheets) made of magnesium alloys with this technique are very effective and require careful selection of process parameters, taking into account the special features of the plasticization of magnesium alloys and the stirring effect in the joint area [4, 22–24].

2. Evaluation of plastic formability of Mg alloys

Various magnesium alloys for plastic deformation have difficulties in carrying out metal forming processes. The evaluation of the plastic formability of magnesium alloys can be conducted by determining the mechanical behavior of samples of tested materials in compression, torsion, and tensile tests. These tests reflect relatively well essential features of the state of stress or deformation in technological processes of metal forming, including extrusion, forging, and rolling, respectively.

To evaluate mechanical behavior of the material in extrusion process, the upsetting test was used to realize plastic deformation under various conditions and to look for adequate their choice for real deformation process. The grades AZ31, AZ61, AZ80, WE43, and magnesium alloys with lithium, as casted ingots and extruded preforms, were used in the research work. In order to study feasibility of these magnesium alloys in extrusion process, the upsetting test of cylindrical specimens was carried out and let to determine flow stress–strain relationships [25–30]. Before upsetting, the specimens were heated in a furnace to established temperature (**Figure 1**).

On the example of AZ31, AZ61, AZ80, WE43 alloys, and magnesium and lithium alloys, the determined flow stress–strain relationships between flow stress and strain for different values of temperature and strain rate allow for adequate adjustment of plastic deformation parameters based on specific relationships between the structure and deformation under conditions of hot compression test. Documented occurrence of two deformation mechanisms: slip and twinning in the presented relationships between plasticity characteristics such as: maximum yield

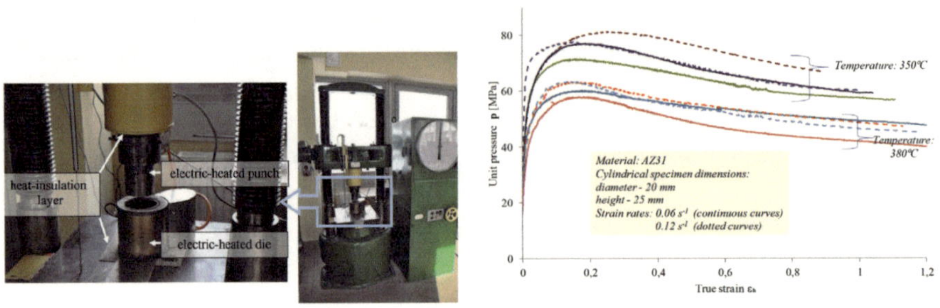

a) b)

Figure 1.
a) Setup of the upsetting/test at high-temperature mounted on 1000 kN hydraulic press, b) flow stress–strain relationships for AZ 31 [28].

stress and strain corresponding to the maximum, and the Zener-Hollomon parameter allows for an appropriate interpretation of the effects of microstructure transformation [31]. In order to prepare technological process, it is necessary to define precisely the plastic properties and microstructure changes of those alloys. Comparison of the plasticity and microstructure of magnesium alloys with from 3 to 8% aluminum content from group Mg-Al-Zn-Mn let to choose proper parameters of the process. On the basis of tensile tests, the plasticity changes were determined at temperature from 150 to 450°C. Conducted compression test at temperature from 250 to 450°C and deformation speed from 0.01 to 10 s^{-1} provided important data concerning the influence of process parameters on flow stress and microstructure changes connected with recrystallization process.

The characteristics of the relationships obtained in the compression test: flow stress σ_p- strain ε for the tested alloys AZ31, AZ61, AZ80, WE43 (**Figure 2**), and magnesium alloys with lithium (**Figure 3**) show the influence of temperature on their course, which allows for an adequate choice of technological parameters for plastic deformation methods with the dominant state of compressive stress (e.g., forging, extrusion).

Tests of magnesium and lithium alloys [32] with lithium content of 2.5, 4.5, 7.5, and 15% of the mass showed very different flow stress–strain relationship characteristics plasticizing from deformation, which results from the fact that alloys with 2.5 and 4.5% lithium form a lithium solid solution in magnesium and have a hexagonal structure. The 7.5% lithium alloy has an.

α + two-phase structure β., where the α-phase with the hexagonal structure is a solid solution of lithium in magnesium and the β phase has a wall structure centered is a solution of magnesium in lithium. The 15% lithium alloy is a solution of magnesium in lithium and has a wall-centered structure. An example of the flow stress–strain relationship for *Mg-7,5 Li* alloy is shown in **Figure 3**.

Alloys that contain more lithium, which is 7.5%, have good formability at temperature of 150°C. The alloy content of 15% lithium demonstrates very good deformability. The shape of flow curves and microstructure of alloys after deformation at elevated temperatures show significant influence of dynamic recrystallization process.

Determination of the plastic formability can be made on the basis of the results of torsion test too [30].

Flow curves for the magnesium alloy AZ31, most commonly used so far, were determined using torsion test at 300, 400, and 450°C at the speed 1 s^{-1} (**Figure 4**), while in the compression test at the temperature of 200, 300, 400, and 450°C at a strain rate of 0.01 and 1 s^{-1}, respectively (**Figure 5**). AZ31 magnesium alloy exhibits

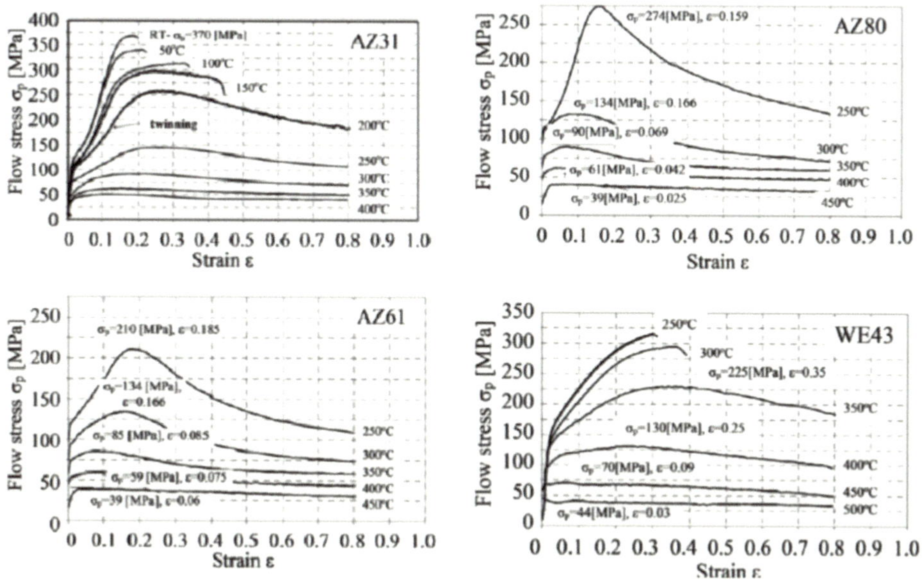

Figure 2.
Flow stress–strain relationships for magnesium alloys AZ31, AZ61, AZ80, and WE43 obtained in compression test [28].

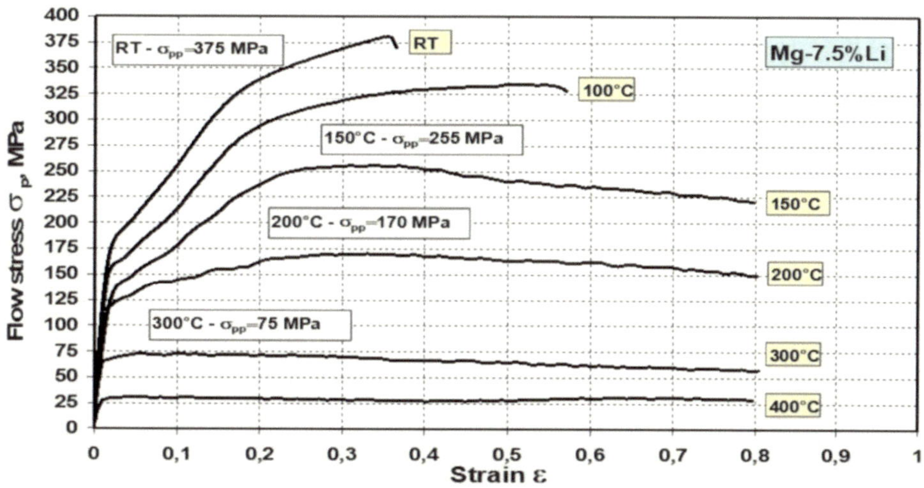

Figure 3.
Flow stress–strain relationships for magnesium alloy Mg-7,5 Li obtained in compression test [28].

an increase in deformability with an increase in the torsional temperature from 1.2 at 300°C to 5 at 450°C (**Figure 4**).

Axial-symmetric compression tests carried out on the Gleeble 3800 simulator with simultaneous "freezing" of the microstructure after deformation by rapid cooling with water, in the temperature range from 200 to 450° C with the strain rate$\dot{\varepsilon}$ = 0.1 s^{-1} and 1,0 s^{-1}, until the deformation ε = 1, 2 (**Figure 5**) allows the assessment of the influence of these factors on the course of the characteristic and its use in the design of plastic forming processes.

The flow curves obtained in the compression test indicate a similar level of the value of flow stress of the alloy for comparable conditions of its deformation in relation to the torsion curves.

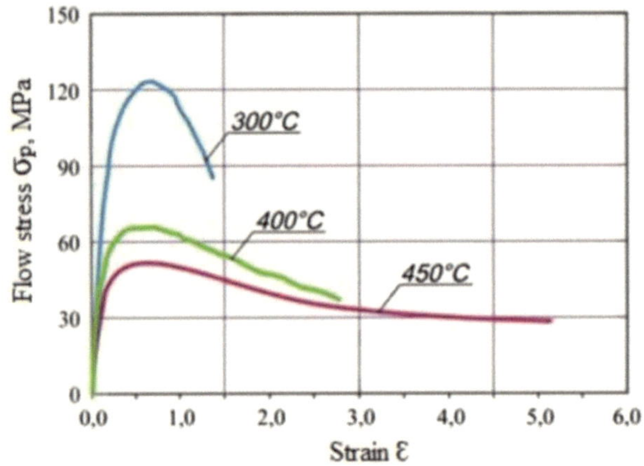

Figure 4.
Flow curve of AZ31 alloy determined in torsion test at temperature of 300, 400, and 450°C with a strain rate 1 s^{-1} [30].

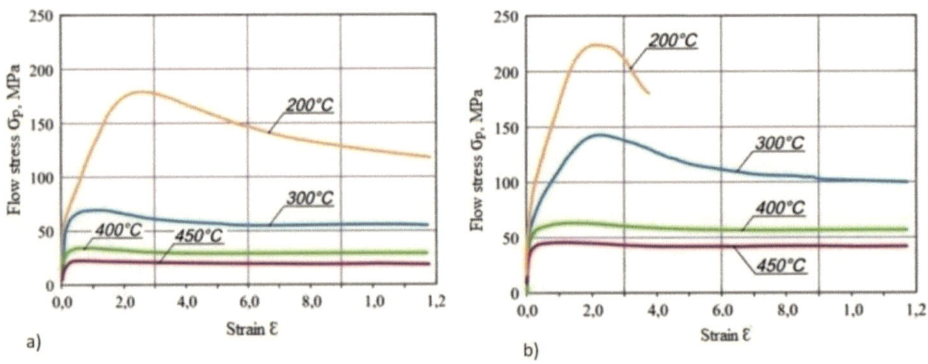

Figure 5.
Flow curve of AZ31 alloy determined in compression test at temperature of 200, 300, 400, and 450°C with a strain rate: a) 0.01 s^{-1}, b) 1 s^{-1} [30].

Performing plastometric tests for a magnesium alloy allows the identification of two types of flow curves. Classical curve—where the dominant mechanism in the microstructure is slip (e.g., **Figure 6a**) and the characteristic curve, where the dominant mechanism in the microstructure is twinning (**Figure 6b**).

The relationship of the maximum yield stress σ_{pp} and strain ε_p as a function of ln Z, where: Z – Zener-Hollomon parameter, $Z = \dot{\varepsilon}\, exp\left(\frac{Q}{RT}\right)$ is shown in **Figures 7** and **8**.

Currently, plastic forming of magnesium alloys is limited to a few basic grades from the group of Mg Al-Zn (AZ21, AZ31, AZ61) and Mg-Zn-Mn (ZM21) alloys. AZ31 magnesium alloy as the most widely used for rolling metal sheets shows good formability under hot-forming conditions. The obtained flow curves depending on the deformation parameters show two different types of the deformation process. For higher temperatures and lower strain rates, the curve follows the classical course of changes in the yield stress. At lower temperatures and higher strain rates, the course of stress changes is different and characteristic for the process based on the twinning mechanism, which was confirmed in structural studies. It has been shown that there is a relationship between the maximum yield stress σ_{pp} and the

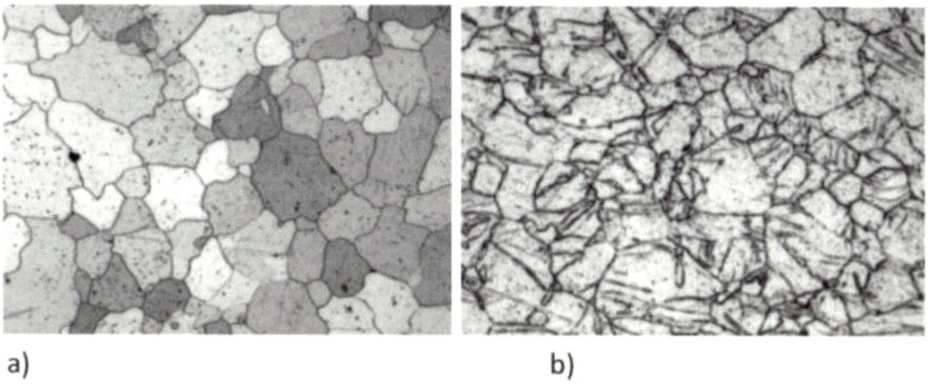

Figure 6.
Microstructure of magnesium alloy AZ31: a) after deformation at temperature of 350°C—Slip domination, b) after deformation at temperature of 250°C—Twinning domination, strain rate 0.1 s^{-1} [30].

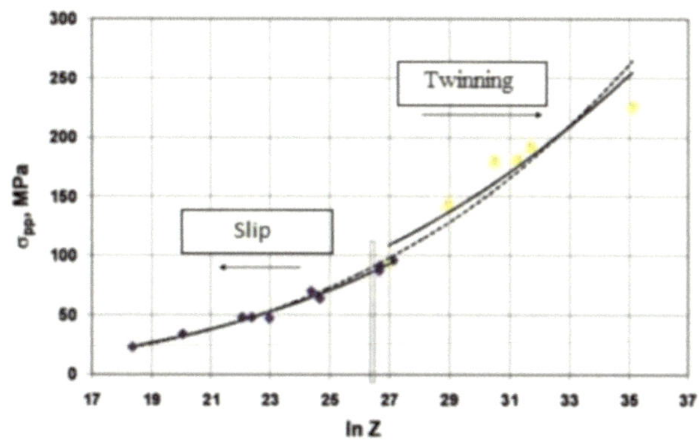

Figure 7.
Maximum yield stress σ_{pp} as a function lnZ [31].

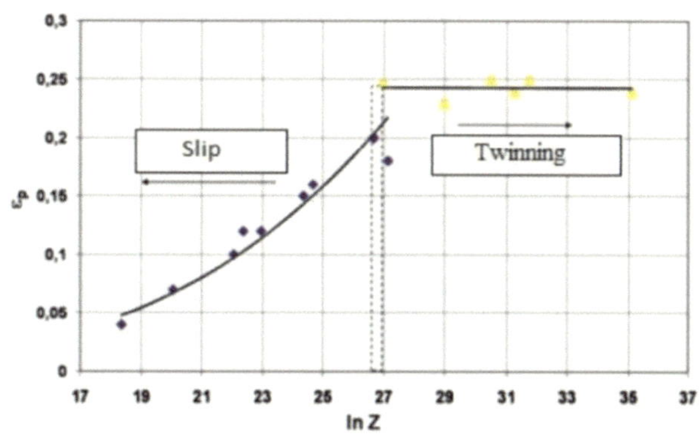

Figure 8.
Deformation ε_p corresponding to the maximum yield stress on the flow curve σ_p as a function lnZ [31].

corresponding strain ε_p, and the Zener-Hollomon parameter (**Figures 7** and **8**). A worse fit occurs for curves where twinning dominates, which changes the shape of the curve [30]. Flow stress–strain curves of alloy AZ31 are characteristic for alloy in which during deformation a mechanism of plastic strain called twinning occurs [33].

The microstructure of AZ31 alloy after deformation by hot compression at the temperature of 200, 300, 400, and 450°C with strain rate of 0.01 s^{-1} and 1 s^{-1}, respectively, was examined, and an example is shown in **Figure 9**. It was observed after the compression test at 200°C for strain =1.2, for both the strain rates used, the microstructure of the primary elongated grains and the ultrafine grains dynamically recrystallized (**Figure 9**). Samples deformed at a lower strain rate are characterized by a greater advancement of the recrystallization process. Recrystallized grains are observed both at the boundaries and within the primary grains.

The microstructure of AZ31 alloy, after deformation at the temperature of 300°C with the speed of 0.01 s^{-1}, consists of fine grains that are dynamically recrystallized. For a higher strain rate, chains of recrystallized grains at the boundaries of the deformed primary grains are observed. Increasing the temperature to 400 and 450°C intensifies the recrystallization processes and grain growth. Few deformation twins are also observed (**Figure 10**).

The presented results of plasticity tests of AZ31 magnesium alloy indicate its good formability during hot deformation. The fine-grained recrystallized microstructure was obtained at the temperature of 300°C for a low strain rate. Increasing the temperature leads to the growth of recrystallized grains. Consequently, the average grain diameter after deformation at 450°C is much higher than before deformation.

Traditionally, the compression test specimens are circular in cross section. Taking into account the geometrical profiles of plastically formed products, e.g., in the

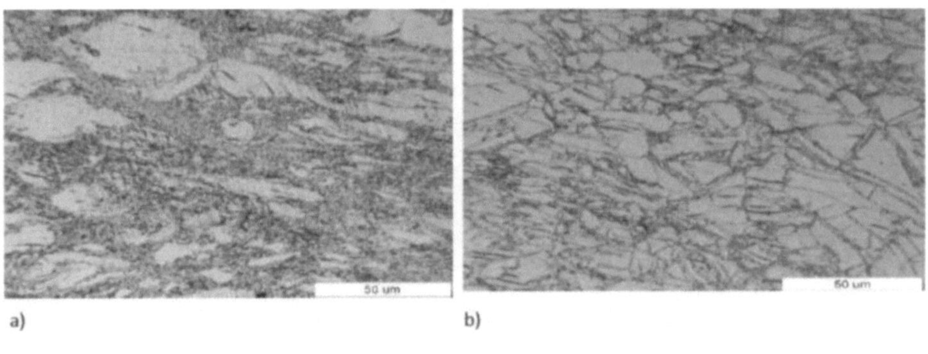

Figure 9.
Microstructure of the AZ31 alloy after compression at temperature 200°C: a) with a strain rate 0.01 s^{-1}*, b) with a strain rate 1* s^{-1} *[31].*

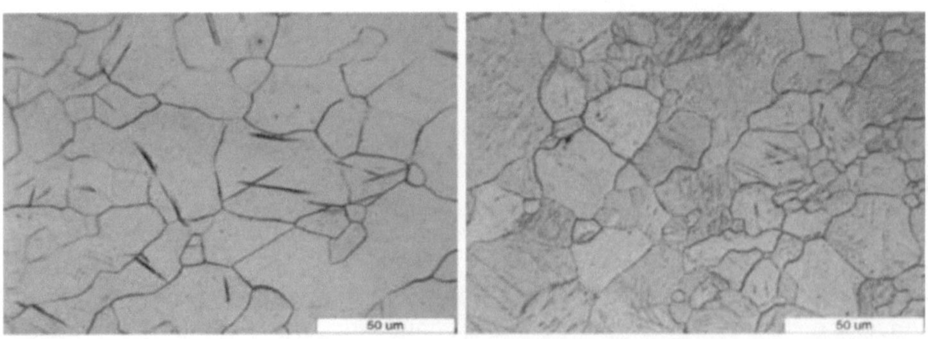

Figure 10.
Microstructure of the AZ31 alloy after compression at temperature 450°C: a) with strain rate 0.01 s^{-1} *, b) with strain rate 1* s^{-1} *[25].*

process of forging or extrusion, the cross-sectional geometries are usually more complex. The analysis of the influence of the geometry of the deformed sample in the non-standard compression test when the cross sections of the deformed sample are not circular (**Figure 11**) shows that the differences are significant in the mechanical behavior of the material (the level and course of the forming force, the change in the geometry of the upset sample).

Taking into account the different geometry of the initial material in the evaluation of the impact of the strain rate and temperature on the forging (upsetting) effect of magnesium alloy specimens (as a material test before designing the forging process) allows for determining the appropriate process parameters.

The results of modeling the upsetting process of magnesium alloys obtained in the form of temperature distributions, stress and strain distribution as well as strain rates provide the basis for determining the conditions of the actual process leading to a product without defects and with high-quality requirements.

The analysis of force courses as a function of displacement during the upsetting process of magnesium alloys showed that the value of the force required for deformation decreases with increasing temperature. The value of the force is strongly influenced by the shape of the upset specimen, including the geometrical parameters (number of corners, the measure of angles, axes of symmetry, and planes of symmetry) (**Figure 12**).

Both in numerical simulation of upsetting and in experimental tests, the values of the force needed to deform individual specimens are convergent. The more complicated the cross-sectional shape, the greater the force needed to deform a given specimen of metallic material.

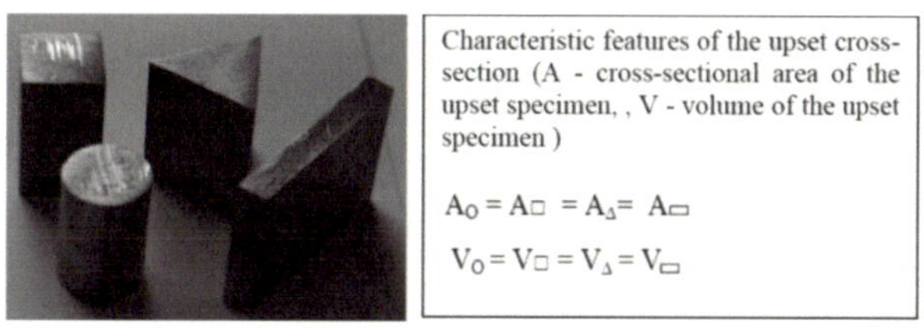

Figure 11.
Upsetting test specimens.

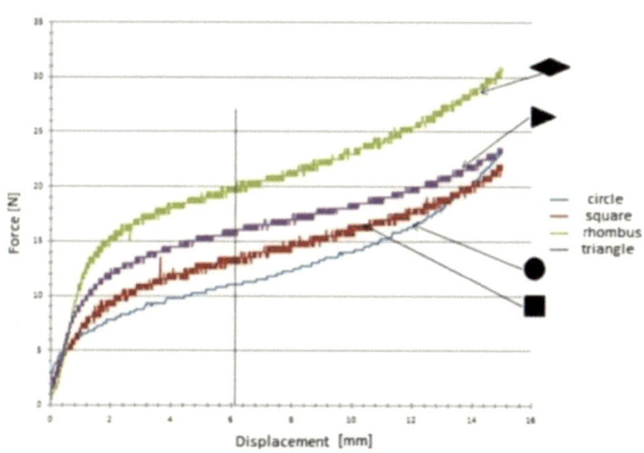

Figure 12.
The effect of different geometry of initial material (shape of the cross section of the sample) on the force of deformation during upsetting test.

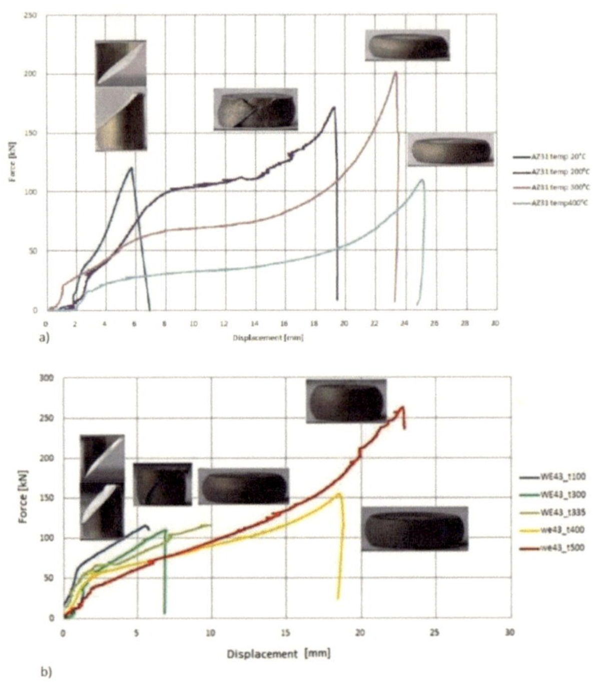

Figure 13.
Influence of temperature on the course of force values during the upsetting of magnesium alloy a) AZ31, b) WE 43.

The results of upsetting test for the forging process demonstrate different mechanical behavior of various Mg alloys: e.g., AZ31 and WE 43, and they are useful to determine plastic formability (**Figure 13a** and **b**).

On the basis of their analysis, it is possible to assess ability to deformation of magnesium alloys on the basis of determining:

- limit deformations, deformations leading to cracking, and force parameters during the process of metal forming.

3. Processes of metal forming: forging, extrusion, KOBO extrusion, rolling, and joining: friction stir welding. Mechanical properties, microstructure, and quality of the final product. Examples of applications

The analysis of magnesium alloys and their possibilities of deformation by plastic forming shows that they are prospective due to the development of a number of new technologies. The purposefulness of work on the development of the plastic forming technique [34] is primarily determined by the better mechanical properties of plastically processed magnesium alloys compared with the cast ones. Now, there are main groups of magnesium alloys available for plastic deformation:

- *Mg-Al-Zn alloys*—The magnesium alloys Mg-Al-Zn are the most popular ones. Four basic alloys AZ31, AZ61, and AZ80 are distinguished. The alloy AZ31 shows relatively low mechanical properties, but it is weldable and perfectly suitable for rolling, stamping, and extrusion. This grade is used to produce sheet metal designed mainly for drawpieces. The alloys AZ61 and AZ80 are characterized by a larger content of alloyed components, and they show more advantageous mechanical properties. The alloy AZ61 is weldable, plastically worked by extrusion and forging methods. The alloy AZ80 demonstrates the best mechanical properties in the group of plastically worked alloys; however, its susceptibility to plastic working is relatively low. It is suitable for making only simple forgings.

- *Mg-Zn alloys*—Two alloys ZM21 and ZC71 are distinguished in this group. The alloy ZM21 includes zinc up to 2% and manganese in the amount of about 1%; it is susceptible to rolling and stamping and characterized by good weldability. A fine-grained structure of average grain size of about 15 μm can be obtained after extrusion. ZC71 is a new magnesium alloy with zinc, copper, and manganese that is characterized by high strength of up to 360 MPa. It can be formed by extrusion and forging methods, and this alloy is weldable.

- *Mg-Zn-Zr alloys*—This group comprises the alloys ZK30, ZK40, and ZK60 that includes 3–6% Zn and 0.4–0.6% Zr. An addition of zirconium leads to an intensive grain refinement. These alloys are characterized by high strength, and they are formed in forging and extrusion processes.

- *Mg-Y-Re-Zr alloys*—The alloys of such a type are formed by the plastic working methods, most often via extrusion. It can be mentioned that the alloy WE43 that, as main component, includes yttrium in the amount of 4.0% and rare earth elements RE, i.e., about 3.5%. After extrusion and heat treatment, the alloy WE43 shows tensile strength Rm equal to 420 MPa, yield point Re = 340 MPa, and elongation A = 15%. This alloy is characterized by good creep resistance at the increased temperatures

- *Mg-Li alloys*—Magnesium-lithium (Mg-Li) alloys thus have distinct advantages over conventional magnesium alloys. However, Mg-Li alloys possess relatively low strength and oxidation resistance. Alloying is well known to be an effective method to improve mechanical and chemical properties. Small editions to Mg-Li system, e.g., aluminum may improve mechanical properties. Mg-Li-Al (LA 143) alloys with high strength and plastic deformability were prepared through a combination of heat treatment and multidirectional forging in a channel die (MDFC). The maximum specific yield strength of the Mg-Li-Al alloy in this study is 263 $kN\cdot m\cdot kg^{-1}$. The limit of reduction during cold rolling of all MDFCed LA143 samples exceeded 99%. The high specific yield strength could be attributed to severe plastic deformation. LA143 alloys with excellent mechanical properties can be prepared by heat treatment and severe plastic deformation [35].

Designing the processes of metal forming of structural elements made of magnesium alloys requires precise determination of the influence of process parameters on the microstructure and consequently on the mechanical properties of the manufactured elements. This is of particular importance when designing products in aviation [36, 37], automotive, medical, and other applications. Plastic forming tests carried out under laboratory and industrial conditions indicate that selected magnesium alloys can be formed especially in the process of rolling, forging, and extrusion. When forming AZ61, AZ80, and WE43 alloys, the temperature range is significantly limited, both at the beginning and the end of the deformation process. Therefore, to carry out plastic forming, especially forging, it is necessary to have devices that enable the process to be carried out in isothermal conditions. For AZ31 alloy, the range of temperatures of good formability is greater due to the greater tendency of this alloy to the recrystallization process.

The beneficial properties of magnesium alloys are obtained thanks to thermoplastic treatment. In magnesium alloys, an intensive process of dynamic recrystallization takes place during plastic deformation, which promotes grain refinement and improvement of mechanical properties. Plastic forming of magnesium and its alloys can be carried out, depending on the content of alloying elements, only in a narrow temperature range.

Examples of the use of magnesium alloys in the formation of products/semi-finished products in the processes of plastic forming and joining processes involving plastic deformation (friction stir welding) show the enormous potential of these materials and the clear benefits of using these technologies in the forming of various types of products.

3.1 Particularities of metal forming processes of Mg alloys

3.1.1 Rolling

Rolling of magnesium alloys is currently limited to a few basic grades from the group of Mg Al-Zn and Mg-Zn-Mn alloys. The new alloys Mg-Th- (Mn or Zr) and Mg-Li-Al are also susceptible to rolling [30, 32, 35, 38, 39].

The process of rolling magnesium alloy products is very expensive and time-consuming. This is due to the necessity to carry out annealing between successive operations. As a result of the current growing interest in sheets of magnesium alloys, the so-called "twin rolls casting" technology has been developed, which reduces the number of rolling and heating operations by casting between rolls and subsequent rolling [32].

Rolling AZ31 alloy with heating the billet in a chamber furnace to the temperature of 470°C for 30 minutes and cooling it in the air to the rolling temperature, in the range from 200 to 450°C, allows for obtaining a product with the required mechanical properties at the level of Rm = 220–265 MPa and A50 = 10–12%.

Examples of the microstructure of AZ31 alloy bands after the hot rolling process are shown in **Figures 14** and **15**. In the microstructure of the specimens rolled with a total draft of 44%, it shows a partially recrystallized structure (**Figure 14a**). The presence of primary grains and recrystallized grains was found (**Figure 14b**) fully recrystallized structure with fine grains. The analysis of the microstructure of the

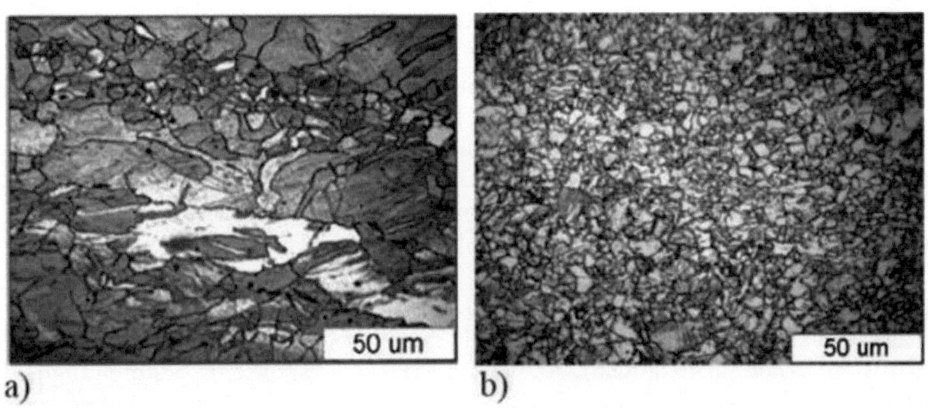

Figure 14.
Microstructure of bands from alloy AZ31 after rolling process by total draft: a) 44%, b) 82% [32].

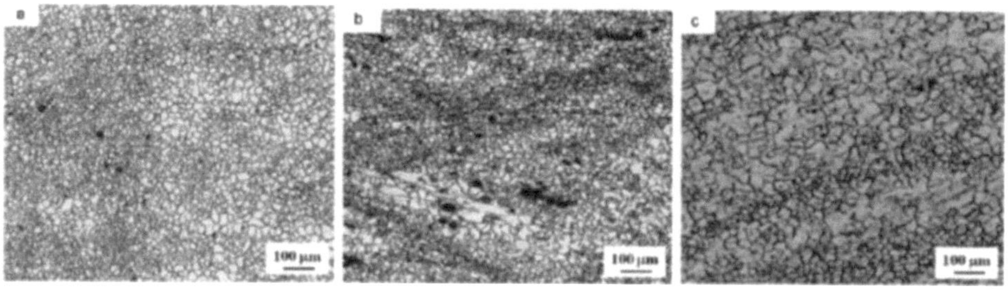

Figure 15.
Microstructure of the AZ31 alloy rolled plate: a) in the surface of the plate, b) in the longitudinal section of the plate, c) in the cross section of the plate [30].

AZ31 alloy bands shows that it is precisely the application of large creases (here 82%) that allows obtaining a fine-grained structure without visible areas of primary grains.

The existing industrial application of magnesium alloys is currently focused on the utilization of semi-finished products such as sheets. The effects of processing parameters and special aspects of the rolling process on the mechanical properties and sheet formability is examined, and recent developments is presented in [37, 40].

3.1.2 Forging

Among the most common forming processes, forging is a promising candidate for the industrial production of magnesium wrought products. The basics of magnesium forging practice are described, and possible problems as well as material properties are presented and discussed in many papers. Several alloy systems containing aluminum, zinc, or rare earth elements as well as biodegradable alloys are evaluated to focus on the process control and processing parameters, from stock material to finished parts including mechanical properties and analysis of microstructure [37, 40, 41].

The final properties of the forgings made of Mg alloys depend on the type of alloy and the string technological process leading to the final product. Most often feet cast magnesium is plastically deformed into semi-finished products, mainly by extrusion or rolling. Usually, forgings are made of semi-finished products in the form of extruded bars or rolled plates, which can be supplied in various conditions depending on the type of heat treatment applied. Forgings after being forged are subjected to precipitation hardening, recrystallization annealing, or stress relief annealing or leaving heat untreated. Each stage of the production line affects the structure and final properties of the product.

Forging as a process of forming the material through multistage deformation for magnesium alloys is a typical process based on hot forming at a narrow temperature range. An example of multistage forging of the airplane wheel hub and the AZ31 magnesium alloy control system lever of the helicopter is shown in **Figure 16** [42–44].

Forging should be carried out on hydraulic or low-speed hydraulic presses. In specialist literature, it is not recommended to use die hammers and high-speed presses due to the cracking of the material during forging. When designing the transverse flow tendency of the magnesium alloys, the rod axis as well as difficult flow in the longitudinal direction should be taken into consideration. Die blanks should be polished to facilitate material flow and avoid surface defects. Free removal of the forging from the blank is possible thanks to forging inclinations equal to 3°, and in some cases even smaller [45, 46].

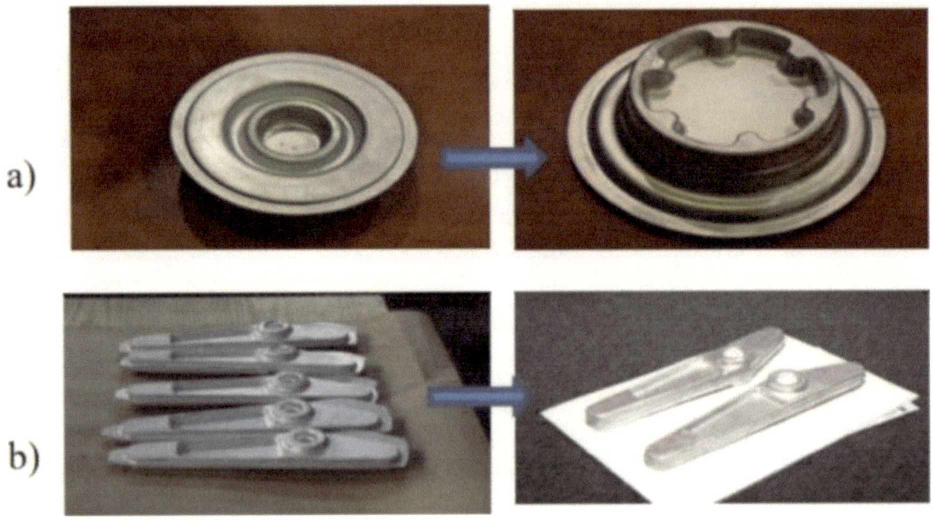

Figure 16.
View of forgings from AZ31 alloy: a) semi-hub of wheel, b) lever after process at beginning temperature of 410°C [28].

A very important process parameter is the temperature of the charge and tools. Magnesium alloys are good heat conductors and quickly cool down in contact with the tools, the more so that during forging, the material deformed over a relatively long time is in contact with a large surface with the die shape. For this reason, the temperature of the tools should be kept slightly below the load. Too much cooling of the billet leads to the formation of cracks. Conversely, too high a temperature also causes cracking due to the occurrence of hot brittleness. Lubrication is applied during the forging operation. Greases based on graphite or molybdenum disulfide are recommended for the forging temperature range of magnesium alloys.

Magnesium alloys are highly strain rate sensitive and exhibit good workability at a narrow forging temperature range. Consequently, parts made of these materials are usually forged with low-speed hydraulic presses, using specially designed tool heating systems in order to ensure near isothermal conditions. This study investigates whether popular magnesium alloys such as Mg-Al-Zn can be forged in forging machines equipped with high-speed forming tools.

Results presented in work [47] have demonstrated that AZ80A is not suitable for forging with either the screw press or the die forging hammer, that AZ61A can be press- and hammer-forged but to a limited extent, and that AZ31B can be subjected to forging in both forging machines analyzed in the study.

Examples of the application of magnesium alloys in aviation structures obtained by forging indicate the effective use of the possibilities of this technology (Examples shown in **Figures 17–22**).

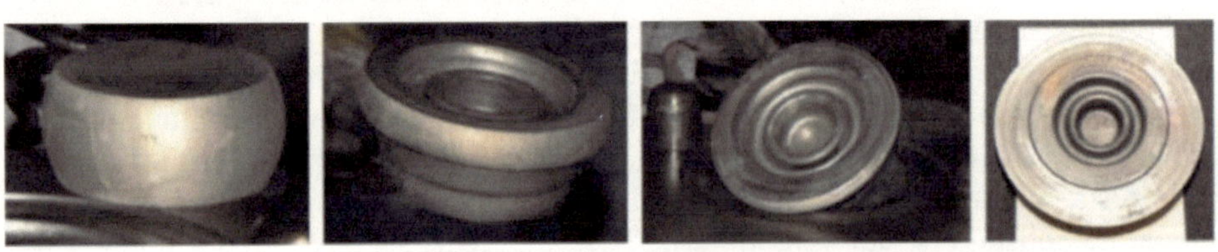

Figure 17.
Stages of forging of AZ31 semihub of wheel, for aeroplane at temperature of 410°C.

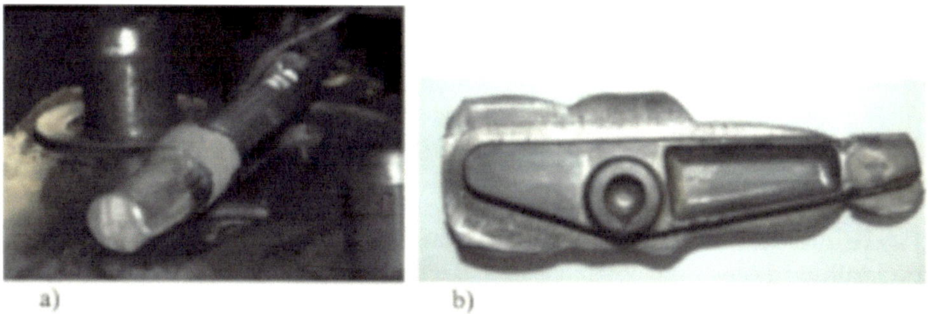

Figure 18.
Stages of forging the helicopter control system lever: a) forging, b) forging AZ31 alloy.

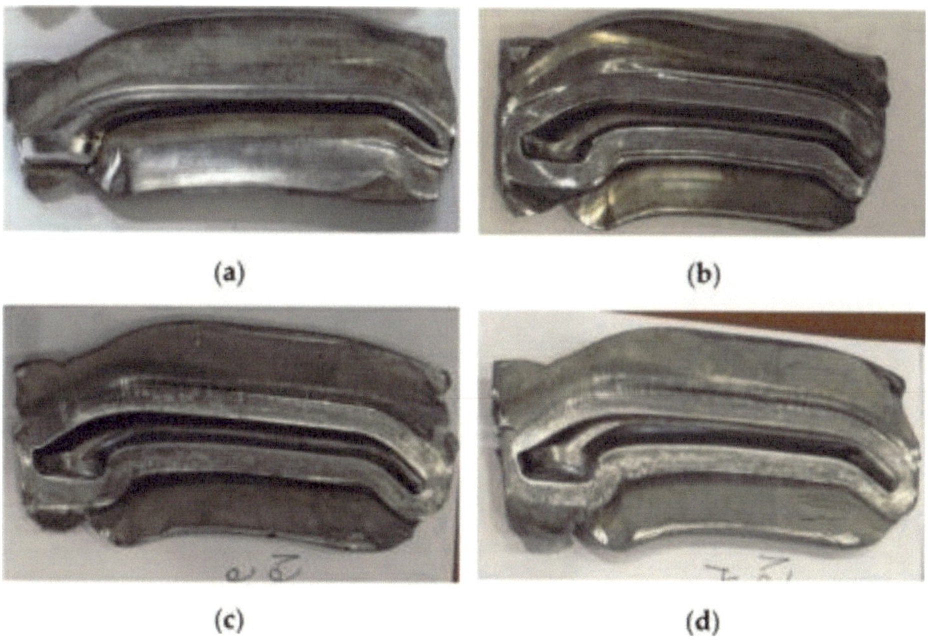

Figure 19.
Drop forgings made in industrial conditions: (a) AZ31B, screw press; (b) AZ31B, forging hammer; (c) AZ61A, screw press; (d) AZ61A, forging hammer [47].

The forging process was carried out on a drop forging hammer. The initial material was ingots heated to the initial forging temperature equal to 350°C and 410°C, followed by upsetting, forging, forging in a die blank. The correct forging was obtained for the alloy annealed at 410°C [41–44].

The results of the research on the forging process in industrial conditions of two selected parts of aircraft structures: i.e., the aircraft wheel hub and the helicopter control system levers showed that the appropriate geometric parameters of these magnesium alloy elements and the determination of the conditions of the hammer forging process allowed for obtaining final products without defects with the required final properties (geometric and mechanical properties).

3.1.3 Extrusion

The conventional process of extrusion of magnesium alloys is carried out at a temperature range from 320 to 450°C, at a speed of 1–25 m/min. The developing method of hydrostatic extrusion allows for plastic deformation at lower temperatures and to obtain greater grain grinding of magnesium alloys [28, 43, 48, 49]. The extrusion process was carried out on a counter-press with a heated container. AZ31, AZ61, AZ80, and WE43 magnesium alloy ingots with a diameter of 100 mm, heated to a temperature of 400°C, were extruded at various speeds from 0.04 to 0.16 m/s and with a different extrusion ratio of 6,25–25. The most favorable effect on the microstructure was observed after billet extrusion with an extrusion ratio of λ = 25 (**Figure 20a–d**). As a result of plastic deformation and recrystallization, fine recrystallized grains were obtained, although in the case of AZ61 and AZ80 alloys, a banded microstructure was observed (**Figure 20b** and **c**).

Microstructure	Mechanical properties	Application
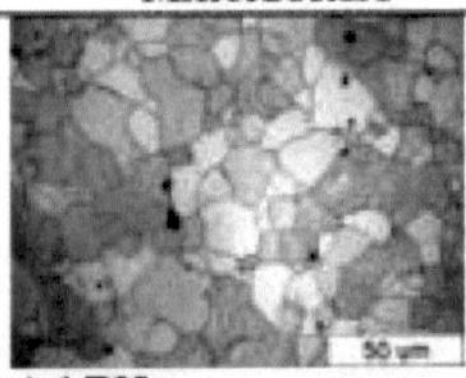 a) AZ31	YS = 170MPa UTS = 264MPa, A_5 = 15%,	sheets, extrusion profiles, forging of complex design, fully weldable alloy
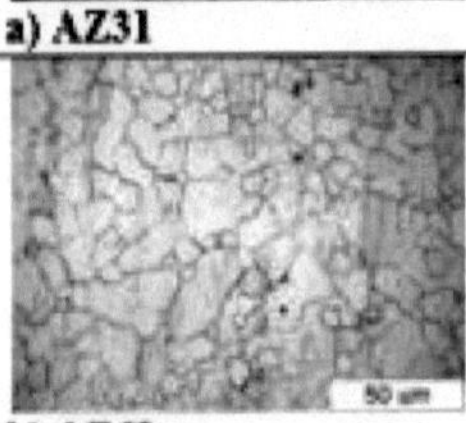 b) AZ61	YS= 240 MPa, UTS =325 MPa, A_5=15,4%	extrusion of complex profiles, forging of complex design, fully weldable alloy
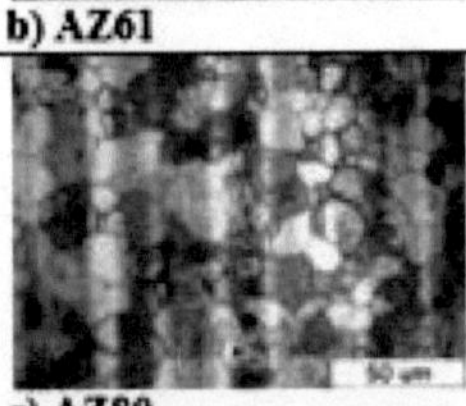 c) AZ80	YS =255 MPa, UTS = 340 MPa, A_5=13,5%	complex shape of extrusion, forging of simple design, fully weldable alloy
 d) WE43	YS =280 MPa, UTS = 330 MPa, A_5= 15 %	high temperature aerospace alloy up to 250°C, complex shape of extrusion, forging of simple design, flat products

Figure 20.
Microstructure and mechanical feature of magnesium alloys after indirect extrusion [28].

Different shapes of cross section of profiles for aviation application were obtained in the way of backward hot extrusion process. Some results of final products are presented in **Figure 21**. It is possible to obtain profiles of complex shape with elements of varied wall thickness and with thin walls. Microstructure of tested alloys in initial condition after extrusion is shown in **Figure 20**. Before deformation the tested alloys AZ31 and AZ61 were characterized by single-phase microstructure of solution α-Mg, whereas in microstructure of alloy AZ80 the presence of intermetallic phase γ-($Mg_{17}Al_{12}$) was found and which was confirmed by prior X-ray tests. The extrusion of profiles with a complex cross-sectional shape and large differences in wall thicknesses (**Figures 21** and **22**) allowed the assessment of high plastic deformation possibilities of these alloys, favorable microstructure, and obtaining very good mechanical properties [28, 49–51].

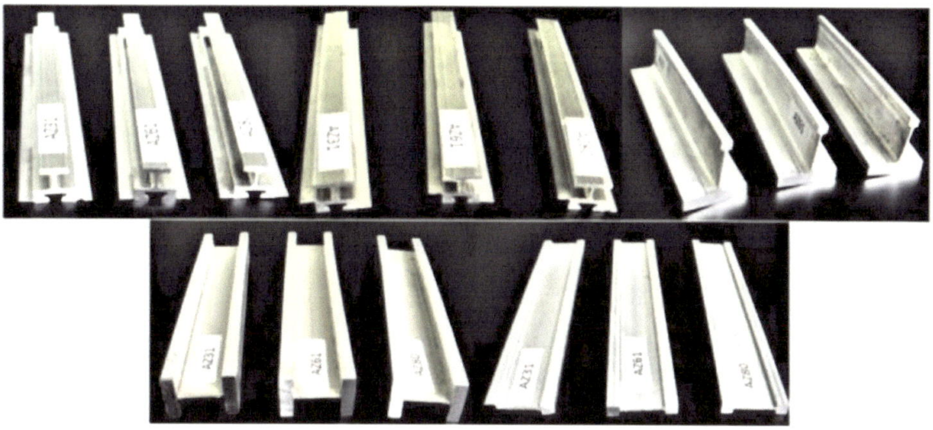

Figure 21.
Complex shape profiles of Mg alloys obtained during indirect extrusion [28].

The measurements of the grain size (acc. to ASTM E 112) and the microhardness HV0.1 of the magnesium alloy on the cross section of the extruded profile in the characteristic areas A, B, C (**Figure 22**) show different values of microhardness and grain size (**Table 1**), which proves the influence of the cross-sectional geometry on the microstructure and properties of the magnesium alloy.

Parameters of the direct and hydrostatic extrusions are presented in **Table 2**. In the field of extrusion of magnesium alloys, the method of hydrostatic extrusion has been developed quite intensively in recent years. Due to favorable thermo-mechanical conditions, the hydrostatic extrusion process can be carried out at lower-temperatures and a greater grain grinding of magnesium alloys [52, 53].

The comparison of microstructures of various magnesium alloys after plastic forming during extrusion with the same extrusion ratio λ = 25 for alloys: AZ31, AZ61, AZ80 WE43 (**Figure 23**) indicates diametrical differences in the final plastic

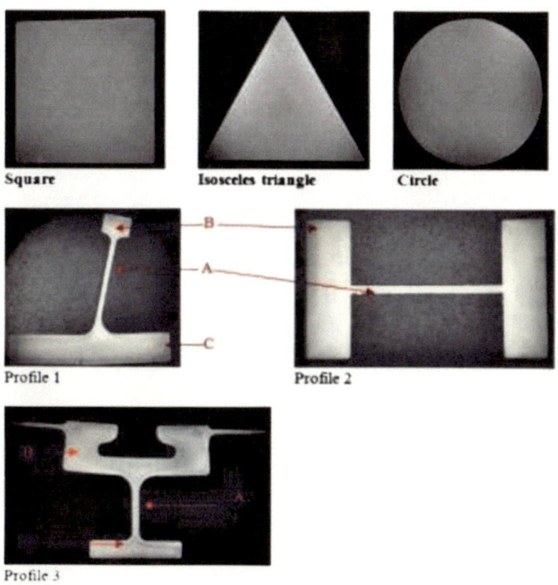

Figure 22.
Grain size measurements on the cross sections of the extruded profiles: Square, Isosceles triangle, circle, and profiles 1, 2, and 3 of complex shape of cross sections and the billet (Ø100 mm) [28].

Alloy	Cross section	zone	Grain size—G, Plate I, ASTM E 112	Microhardness HV0.1
AZ31	Billet– Ø 100 mm		7.5	65
	Square		7.5	58
	Isosceles triangle		8.5	65
			7.5	56
	Profile 1	A	10.5	
		B	10.5	
		C	10	
	Profile 2	A	10	
		B	9.5	
	Profile 3	A	10	
		B	9.5	
		C	10	
AZ61	Billet– Ø 100 mm		8.5	65
	Square		9	56
	Isosceles triangle		9.5	62
			9.5	63
	Profile 1	A	9	
		B	10.5	
		C	9,5	
	Profile 2	A	9	
		B	10	
	Profile 3	A	9	
		B	9	
		C	9	
AZ80	Billet– Ø 100 mm		7.5	60
	Square		9.5	58
	Isosceles triangle		9.5	60
			9.5	59
	Profile 1	A	9.5	
		B	9.5	
		C	9.5	
	Profile 2	A	9	
		B	10	
	Profile 3	A	9	
		B	9	
		C	9	

Table 1.
Grain size and hardness HV0.1 of magnesium alloys profiles after extrusion.

deformation effect and let to choose adequate alloy for requirements of given possible application.

Research work using magnesium alloys AZ31, AZ61, AZ80, WE 43, and Mg alloy with Li for production of thin-walled profiles showed the potential of the backward extrusion process of magnesium alloys and KOBO extrusion process to apply them for aeronautical applications. It gives the possibility for indicating the best materials and parameters of the process to use, e.g., bridge dies to produce different types of complex shape of cross sections, which are the interest of aviation industry in light extruded products in order to save energy, costs, etc., by reducing aircraft weight.

Mg alloy	Direct extrusion temperature [°C]	Direct extrusion Speed [m/min]	Hydrostatic extrusion temperature [°C]	Hydrostatic extrusion speed [m/min]
AZ31 (Mg-Al-Zn)	320–380	8–15	190	45
AZ61 (Mg-Al-Zn)	320–380	2–4	200	10
ZM21(Mg-Zn-Mn)	320–380	3÷5	200	30

Table 2.
Parameters of the direct and hydrostatic extrusion.

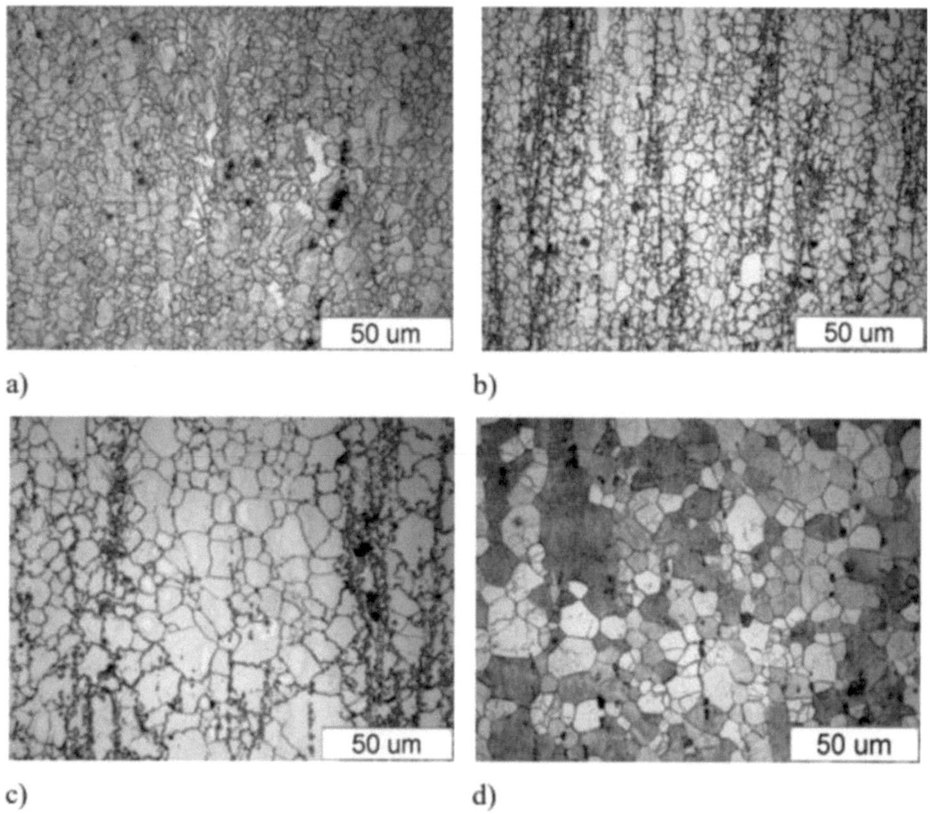

Figure 23.
Microstructures of magnesium alloys after extrusion with the degree of = 25: a – AZ31, b – AZ61, c – AZ80, d – WE43 λ plastic forming.

3.1.4 KOBO extrusion

Processes of severe plastic deformation (SPD) are defined as metal forming processes in which a very large plastic strain is imposed on a bulk process in order to make an ultrafine-grained metal [54]. Possibility to influence on the microstructure and the mechanical properties by heat treatment is also important to plan an SPD process [52, 55]. The KOBO extrusion process as SPD process is an unconventional extrusion method based on the idea of cyclical deformation path change during the process and localized plastic flow [53, 56]. The deformation path change is carried out by setting the die into an oscillating rotary motion around its axis. The method combines the extrusion of the material with an additional plastic deformation caused by reversible torsion of die. Being induced this way super plastic mode of deformation makes possible to deform metals with very high deformation (SPD process) at low temperature (room temperature). Hence, despite very low load capacity of the press, the methods allow for the variety of metallic products to be extruded at room temperature from the billet, with size and dimensions with value of extrusion ratio λ up to several hundreds. The example for KOBO extrusion of MgLi4 is presented in **Figure 24**.

Superplastic behavior of a metals under such deformation conditions is proven also in exact filling of a die opening, regardless how complicated it is and mode of extrusion forward or sideway. The most important advantage of the KOBO method is the ability to deform metals with a very high λ coefficient (extrusion ratio) [53, 57–59] in the "cold" process (without preheating the billet or tooling), regardless of the hardening state of the initial material. The use of the KOBO method allows for reducing the work of deformation, so the extrusion forces are incomparably lower than in conventional extrusion processes. The examples of measured parameters of KOBO process for magnesium alloys AZ31 and WE43 are shown in **Figure 25**.

KOBO extrusion is a process carried out in a system of concurrent extrusion with an oscillating matrix. In turn, the possibility of forming the products of a specific cross-sectional shape, as in conventional extrusion processes, makes it possible to obtain products, not only specimens, with very fine grains and favorable mechanical properties, which qualify the KOBO method for potential industrial implementations. The significant reduction of the extrusion load and the lack of the need for preheating make the method very economically attractive. Another, extremely attractive feature of the KOBO method is the possibility of low-temperature consolidation of postprocessing chips [28, 53, 57–64] and obtaining a solid product as a result of plastic deformation in the process. The deformation effects of magnesium alloys, both in the form of a solid billet and in the form of consolidated postprocessing chips, are shown in **Figure 26**.

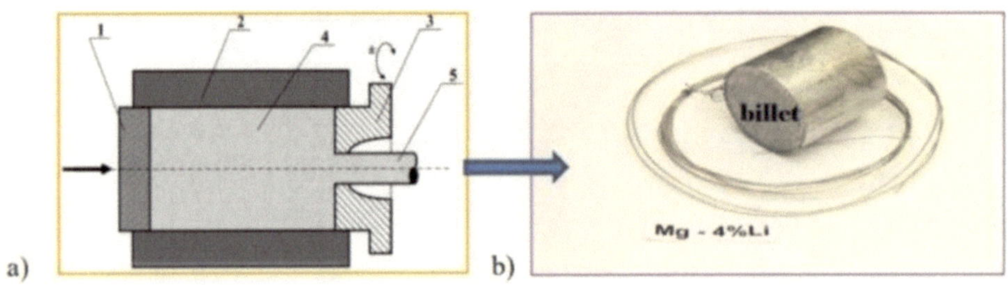

Figure 24.
a) Scheme of KOBO extrusion process, b) extruded wire of Mg Li4 alloy, extrusion ratio λ = 10,000.

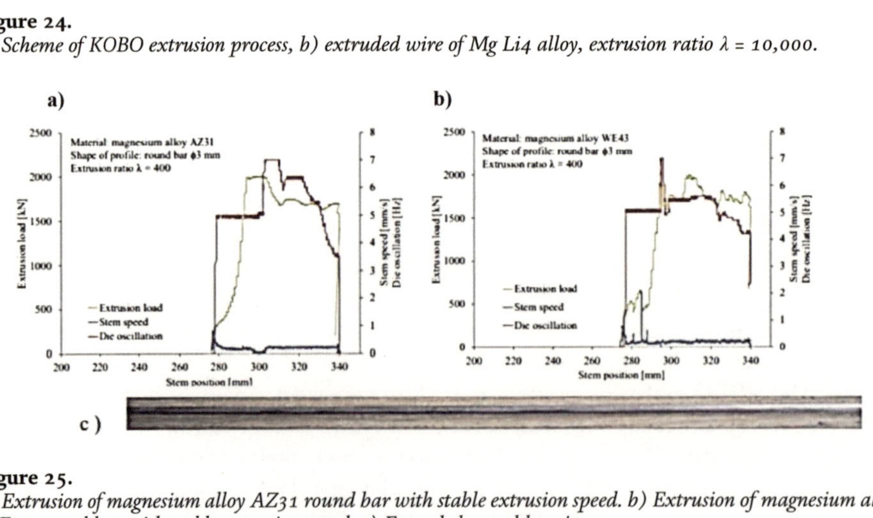

Figure 25.
a) Extrusion of magnesium alloy AZ31 round bar with stable extrusion speed. b) Extrusion of magnesium alloy WE43 round bar with stable extrusion speed. c) Extruded round bar ϕ3 mm.

a) b) c)

Figure 26.
Extrusion of metal chips in the KOBO process: a) a chip before and after compaction of chips into billet, b) an extruded monolithic profile, c) extrusion of solid profile of complex shape of cross section.

Low temperature—above 200°C in KOBO extrusion technology of magnesium alloys, possibility to control the final properties of the extruded profiles let to maintain a fine grain structure of the alloys [65]. It comes from the number of KOBO process parameters, such as amplitude and frequency of die oscillations. It is possible to use the backward extrusion and KOBO processes for bulk metals and chips. To determine proper parameters of extrusion process, it is necessary to know exact information on initial features of given magnesium alloy—its macro and microstructure, mechanical properties, and final results of transformation of the internal structure under conditions of plastic deformation.

3.1.5 Joining. Friction Stir Welding

Joining. Friction Stir Welding FSW is solid-state joining process under the influence of frictional heat. There is no melting involved in the process unlike conventional fusion welding process. The FSW process was originally invented in institute TWI in the United Kingdom in 1991 [65–67]. The principle of utilizing the frictional heat between a rotating tool and the two joining metal interfaces is shown in **Figure 27**.

The rotating tool serves two basic purposes: heating of the workpiece materials due to friction and plastic deformation and stir movement and containment of materials to produce joint. Rotating tool with a specially designed pin and shoulder is inserted into the abutting edges of sheets or plates to be joined and subsequently traversed along the joint line. Advancing and retreating side orientations require knowledge of the tool rotation and travel directions. It is especially important when FSW process is designed for given metallic material. Magnesium alloys require special attention from the point of view of adequate parameters choice, especially for joining very thin sheets, because FSW process is very sensitive to the technological parameters of welding. From the point of view of plasticization of metals, the FSW process is very complex. The rotating tool in the first phase of the process supplies more and more heat, then the welded material is strongly deformed plastically. The material, in contact with the pin of the moving tool, experiences a state of stress and deformation comparable to that in extrusion and forging. The last stage is cooling of the weld [67]. It is important to know the influence of the selected technological and geometrical parameters of the process on the plasticization of the materials to be joined.

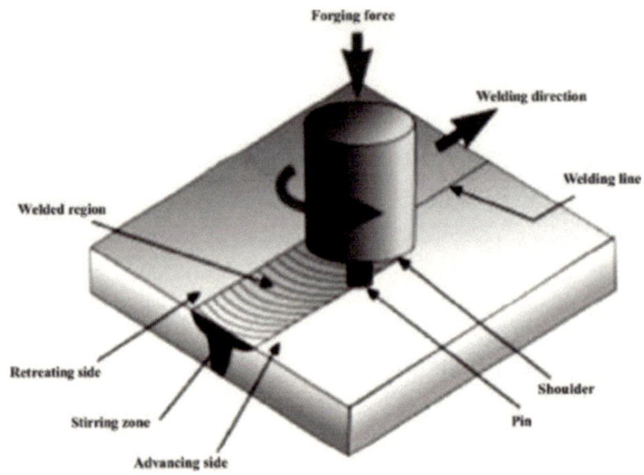

Figure 27.
Scheme of linear Friction stir welding process.

Assessment of joint quality on the basis of tests of the mechanical properties of the obtained joints (static strength, fatigue strength, microhardness), microstructure tests in the area of weld, and parent material as well as the measurement of forces acting on the tool as a material response to the load resulting from the adopted process conditions. As a result of the conducted research on linear, FSW stir-mixed butt welding of thin sheets with a thickness of 0.5 mm made of AZ31B magnesium alloy, the basic goal was achieved, which was to obtain a solid, durable, free from defects connection of elements. When welding 0.5 mm thick sheets, the variable parameters were the rotational speed and the tool feed. The friction stir welding process was carried out with the use of magnesium alloy sheets, including AZ31B, 0.5 mm thick. The FSW joining process was carried out as a butt joint. An example of a connection is shown in **Figure 28**.

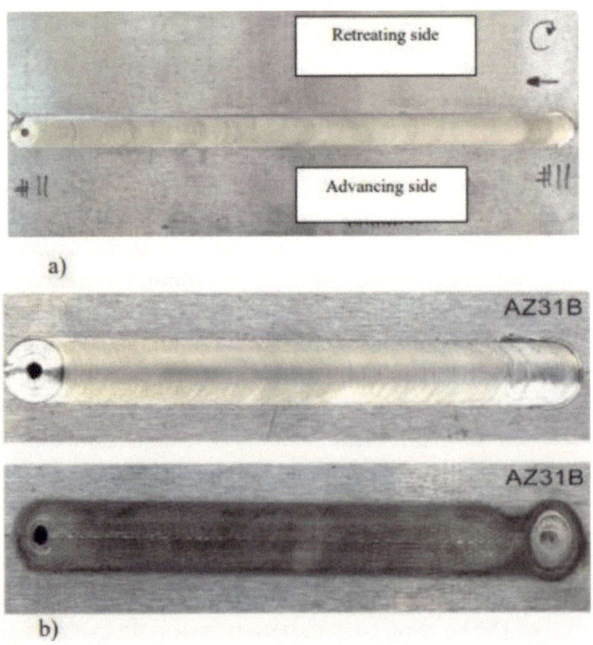

Figure 28.
FSW joints: a) view of joint with advancing and retreating sides, b) FSW joint of the AZ31B magnesium alloys, a) face and ridge of joint FSW joint of AZ31 B sheets of 0.5 mm of thickness, l = 180 mm.

Due to the difficult conditions for the implementation of the process, an important issue is the selection of an appropriate tool with the appropriate geometry and kind of material of the tool. During FSW welding of magnesium sheets a cemented carbide tool was used. Taking into account the interdependence of mechanical and structural effects in the FSW process, special attention should also be paid to the analysis of the microstructure in the joint area, identification of characteristic zones as a result of the transformation of the microstructure resulting from the plasticization of the joined materials and the generated heat due to friction and plastic deformation, heat dissipation, among others in contact with the tool and tooling and the overall heat balance resulting in the final state of the joint. These are issues for thin sheets, especially for light metals (aluminum alloys, magnesium alloys) [68, 69]. **Figure 29** shows the FSW weld microstructure of a sheet with a thickness of 0.5 mm of AZ31B alloy. The presented sample was made with the welding speed of 80 mm/min and the tool rotation speed of 2000 rpm. It is difficult to identify the zones characteristic for the FSW joint in the case of welding very thin sheets. In this case, the locations of areas such as the heat affected zone, the thermomechanical interaction zone, and the weld core were estimated based on the dimensions of the tool. The area of direct contact of the rim of the tool resistance is characterized by fragmented grain due to high plastic deformation. Equal-axis grain was also observed in the HAZ (heat-affected zone) and TMAZ (thermomechanical-affected zone) as shown in **Figure 28**.

The presence of equiaxed grains in the HAZ and TMAZ zone indicates that the material has fully recrystallized during the FSW process. The dynamics of recrystallization affects the grain size. Magnesium alloys are more prone to dynamic recrystallization than aluminum alloys because the Mg recrystallization temperature is approximately 523 K—lower than that for aluminum alloys. The factor that inhibits grain growth as a result of high temperature is also the cooling rate. In ultrathin materials (sheet thickness equal to or less than 0.5 mm), it is relatively high—the material cools down quickly, which is noticeable already during the process. The joint shown in **Figure 30** does not contain any crater defects or discontinuities.

The obtained results of the FSW joining process of magnesium alloys showed that the process of linear friction stir welding (FSW) is a favorable method of joining difficult-to-weld magnesium alloys [70–73]. Appropriate selection of parameters allows for obtaining joints free from defects. The assessment of the quality of the joints made on the basis of: the results of the joint microstructure analysis, measurements of the microhardness of the joint and the base material as well as the static tensile test of the weld material allowed for an unambiguous determination of the joint effectiveness. Proper selection of technological and geometric parameters allows for obtaining a connection with approximately 90%

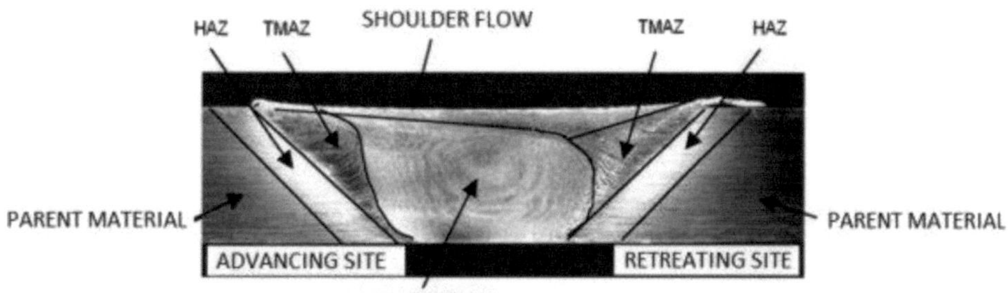

Figure 29.
Typical scheme of zones of butt FSW joint (SZ – Stir zone/nugget zone, TMAZ – Thermomechanically affected zone/HAZ – Heat-affected zone/BM – Base material/parent material, RS – Retreating side, AS – Advancing side).

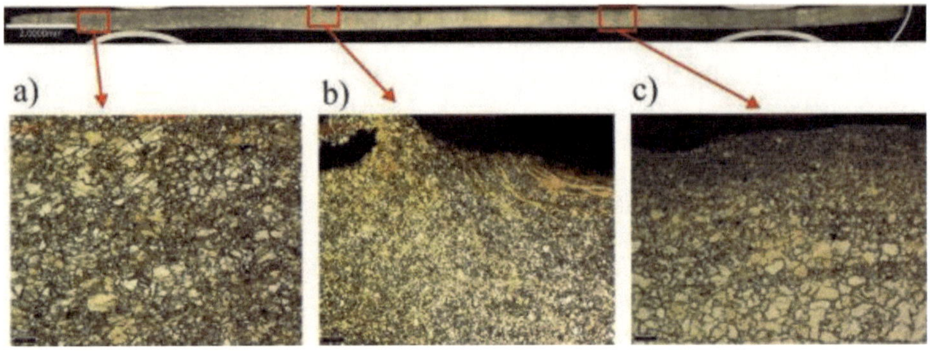

Figure 30.
Typical microstructures of an FSW joint made at different rotational speed and welding speed. a) Microscopic structure of FSW welded AZ31B of 0.5 mm in thickness. View of base material BM. b) Microscopic structure of FSW welded AZ31B of 0.5 mm in thickness. View of HAZ and TMAZ zone from the retreating side. c) Microscopic structure of FSW welded AZ31B of 0.5 mm in thickness. View of stir zone SZ and TMAZ of advancing side.

efficiency in relation to the parent material, determined on the basis of a static tensile test. During the tensile test, all joints broke in the thermomechanical impact zone on the retreating side. This is due to the resulting structural notch at the boundary of the weld nucleus and the thermomechanical influence zone. When welding 0.5 mm thick sheets of Mg AZ31B alloy, the maximum force acting on the tool in the Z axis was 4.5 kN. The values of the forces acting on the tool in the radial direction range from 40 to 100 N. After making a total of approximately 20 m of linear FSW weld, the tungsten carbide tool was not worn according to visual assessment. The FSW process is an excellent alternative to riveting or conventional welding. It does not require the use of an additional connector, which contributes to reduce the weight of the structure. Friction stir welding of the FSW joint allows for the creation of a high-quality, durable, defect-free joint with high mechanical properties (static and dynamic tests) and favorable internal structure of the AZ31B alloy sheet material.

The average obtained connection efficiency with appropriately selected input parameters of the process ranges from 85 to 95% compared with the parent material. These values are fully acceptable to the aviation (min. 80%) and automotive (min. 70%) industries. The analysis of the variability of forces generated during the FSW process requires knowledge of material properties and process conditions. It is necessary to adequately select the process parameters to the type of alloy and thickness of the joined elements (**Figure 31**).

The estimated value of the FSW process temperature should be approximately 0.8 solidus temperature. In this case, for alloy AZ31B, solidus temperature is 605°C, so temperature for FSW process is 430°C. AZ31 alloy with aluminum as the main alloy additive is very brittle at room temperature (only one slip plane (0001)). Only when the temperature of 200°C is exceeded, other slip planes are activated, which facilitates further plastic deformation. Reaching the temperature of 200°C is associated with the phenomenon of recrystallization. Achieving these conditions during welding is possible while providing the rotating and plunging tool with a sufficiently long time stoppage. This treatment is necessary to heat the material due to the pressure and friction of the rotating tool (dwell time). The recorded force courses may indicate the most favorable time to achieve the plasticizing effect and achieve the appropriate plasticization conditions of the material over 450 MPa, which exceeds the value of the yield stress for AZ31B alloy. A sudden increase in radial forces acting on the tool was noticed. Then the welding stage begins, where the stabilization of both axial and radial forces was observed. Very

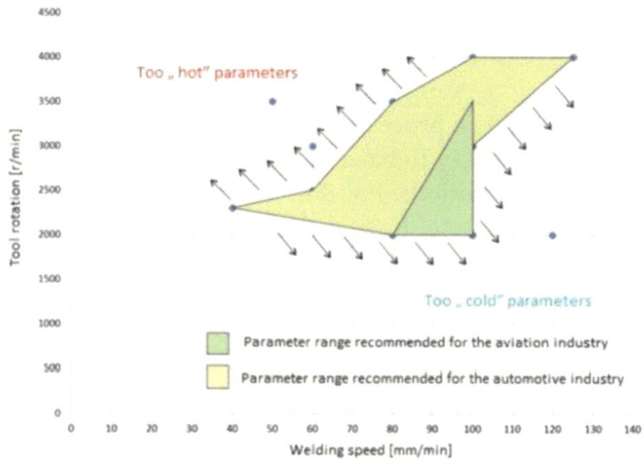

Figure 31.
Matrix of technological parameters for joining AZ31B alloy sheets of 0.5 mm thickness [68].

good results of joining thin sheets of the AZ31 B alloy [64–69, 74, 75] prove the purposeful use of friction stir welding to join sheets of various thicknesses. Butt joints with smooth surface, without voids and flash, can be obtained by cylindrical flat shoulder and pin tool made from tungsten carbide. Tensile tests revealed a durability increase up to 90% compared with BM, which is thought to be mainly attributed to the preferred basal orientation and the activation of the extension twins. The SZ and TMAZ experienced full dynamic recrystallization and thus consisted predominantly of equiaxed grains. The grain size in SZ increased with increasing heat input.

During the FSW, the process must be carried out at the temperature preferably higher than the recrystallization temperature of the base material BM to be joined for the dynamic recrystallization to take place in the SZ. With increasing tool rotational speed or decreasing welding speed supplied more heat energy and generated a higher temperature in stir zone SZ. This led to a weaker or more random texture stemming from the occurrence of more complete dynamic recrystallization. After the FSW of AZ31B alloy of 0.5 mm in thickness, the lowest hardness occurred at the center of SZ through the HAZ and TMAZ of the welded joints; however, the differences are minor.

The welding speed and rotational speed had a strong effect on the UTS (ultimate tensile strength). When choosing the technological parameters for the process, tool feed rate played an important role compared with a tool rotation speed. The plastic flow in the welded regions is also observed with uniform grain orientation.

4. Summary

The development of plastic forming processes as well as joining processes, such as FSW, is primarily determined by better mechanical properties of plastically processed magnesium alloys compared with castings. Designing the technology of plastic forming of structural elements made of magnesium alloys requires precise determination of the influence of the process parameters on the microstructure and consequently on the mechanical properties of the manufactured elements. This is of particular importance when designing products made of magnesium alloys for structural elements for the aviation industry. The selected results of plastic forming research carried out in material tests and technological processes, under laboratory or industrial conditions presented in this chapter, indicate that selected magnesium alloys can be formed by plastic forming methods, especially by rolling, extrusion, and KOBO extrusion and forging. When forming AZ61, AZ80, and WE43 alloys, the temperature range is significantly limited, both at the beginning and the end of the deformation process. Therefore, to carry out plastic forming, especially forging, it is necessary to have devices that enable the process to be carried out in isothermal conditions. For AZ31 alloy, the range of temperatures of good formability is greater due to the greater tendency of this alloy to the recrystallization process. Thus, it is possible to manufacture AZ31 alloy products on conventional devices, but the obtained mechanical properties are less favorable than other magnesium alloys.

Magnesium-lithium alloys deserve attention due to their plastic formability, mechanical properties, and low specific weight, which makes them very attractive wherever lightweight and durable structures are desired. Elements manufactured by plastic processing of magnesium alloys and new technologies involving plastic deformation, including joint friction welding (FSW) technologies, are currently successfully implemented in various sectors of the economy.

Magnesium and its alloys are mainly used as a construction material (most often in the form of castings from magnesium alloys, but also plastically formed products), as an alloy additive to aluminum alloys, and for desulfurization of iron and steel. Due to the low specific mass and high relative strength, magnesium alloys in the form of castings or plastically formed products are used in such industries such as: aviation and aerospace, for the production of aircraft and rocket parts, including engine parts, gearbox components, hinges, fuel tanks, wing elements; automotive, for the production of, among others car rims, various types of housings, engine blocks, steering wheels, seat frames, windows and doors, body parts; sports and recreational, for bicycle parts and elements of various sports equipment articles; electronic, mainly for the production of various types of electronic equipment housings; medical, for strengthening elements in bone fractures. Examples of applications of magnesium alloys in the aviation industry include: Rolls Royce gear housing made of ZRE1 alloy Pratt & Whitney Canada PW535 engine housing made of ZE41 alloy, helicopter parts. In the automotive sector, the examples are steering wheel, boot lid; manufacturer GM, BMW engine block; cross section of the outer layer made of Mg alloy revealing the inner layer of Al alloy.

Now it is time for the successful application of magnesium-based materials. It is particularly important to promote exchange of information and discussion in which development trends and application potential in different fields such as the automotive industry and communication technology in an interdisciplinary framework [37, 72, 73].

Acknowledgements

The author gratefully acknowledges the collaboration and discussions with.

Prof. Eugeniusz. Hadasik and Prof. Andrzej Gontarz, who provided support and information on magnesium research and application.

Financial support of Structural Funds in the Operational Program–Innovative Economy (IE OP) financed from the European Regional Development Fund–Project "Modern material technologies in aerospace industry," Nr POIG.01.01.02-00-015/08-00 is gratefully acknowledged.

DOI: http://dx.doi.org/10.5772/intechopen.101034

Author details

Romana Ewa Śliwa
Rzeszow University of Technology, Rzeszów, Poland

*Address all correspondence to: rsliwa@prz.edu.pl

References

[1] www.simflight.com [Accessed: March 31, 2010]

[2] www.fas.org [Accessed: March 31, 2010]

[3] Kawalla R. Magnez i stopy magnezu. In: Hadasik E, editor. Przetwórstwo metali. Plastyczność a struktura. Gliwice: Politechniki Śląskiej; 2006

[4] Campbell FC. Manufacturing Technology for Aerospace Structural Materials. Amsterdam, Boston, London, New York, Oxford, Paris, San Diego, Singapore, Sydney, Tokyo: Elsevier; 2006

[5] Saravanan A, Suresh P, Sudharsan G. Magnesium alloys: a review of applications. Materiali in tehnologije / Materials and Technology. 2019;**53**: 881-890

[6] Friedrich HE, Mordike BL. Magnesium Technology. Berlin, Heidelberg: Springer; 2006. pp. 219-430

[7] Carou D, Davim JP. Machining of Light Alloys. Taylor & Francis Group: CRC Press; 2019

[8] Tiwari BL, Bommarito JJ. Essential Readings in Magnesium Technology. Willey; 2014. pp. 573-579

[9] Blanchard PJ, Hill DI, Bretz GZ, McCune RC. Essential Readings in Magnesium Technology. Willey; 2014. pp. 593-798

[10] Bender S, Gollner J, Heyn A, Blawert C, Srinivasan PB. Fundamentaks of Magnesium Alloy Metalllurgy. Woodhead Publishing; 2013. pp. 232-285

[11] Perguleryuz MO, Krainer KU, Kaya AAR. Fudamentals of Magnesium Alloy Metallurgy. Oxford, Cambridge, Philadephia, New Delhi: Woodhead Publishing; 2013

[12] Rosochowski A. Severe Plastic Deformation Technology. Dunbeath, Caithness, Scotland, UK: Whittles Publishing; 2017

[13] Bochniak W. Teoretyczne i praktyczne aspekty plastycznego kształowania metali: metoda KoBo. Kraków: Wydawnictwa AGH; 2009

[14] Lewandowska M. Wytwarzanie materiałów nanokrystalicznych metodą wyciskania hydrostatycznego. Obróbka Plastyczna Metali/Metal Forming, T.17. 2006;**4**:9-13

[15] Valiev RZ, Estrin Y, Horita Z, Langdon TG, Zechetbauer MJ, Zhu YT. Producing bulk ultrafine-grained materials by severe plastic deformation. JOM. 2006;**58**(4):33-39. DOI: 10.1007/s11837-006-0213-7

[16] Huana L, Jiapenga S, Jinghua B. Potential of multi-pass ECAP on improving the mechanical properties of a high-calcium-content Mg-Al-Ca-Mn alloy. Journal of Magnesium and Alloys. 2019;**7**(4):617-627

[17] Bohlen J, Letzig D, Kainer KU, Sillekens WH, Vet PJ. Hydrostatic extrusion process for the efficient production of magnesium structural components. The MAGNEXTRUSCO project 12th Magnesium Automotive and End User Seminar

[18] Reed-Hill RE, Robertson WD. Deformation of magnesium single crystal by nonbasel slip. Journal of Metals. 1957;**9**:496-502

[19] Bohlen J, Letzig D, Kainer KU. New perspectives for wrought magnesium alloys. Materials Science Forum. 2007; **546-549**:1-10

[20] Kiełbus A, Kuc D, Rzychoń T. Stopy magnezu – mikrostruktura, właściwości i zastosowanie, Nowoczesne materiały

metaliczne – teraźniejszość i przyszłość, Katowice. 2009. pp. 61-97

[21] Bajor T, Dyja H, Laber K. Physical modelling of metal forming of bars made of magnesium alloys (AZ61). Archives of Metallurgy and Materials. 2015;**60**:3007

[22] Johnson R, Threaadgill P. Friction stir welding of magnesium alloys. In: Mathaudhu SN, Luo AA, Neelameggham NR, Nyberg EA, Sillekens WH, HI, editors. Essential Readings in Magnesium Technology. Willey; 2016. pp. 487-492

[23] Zettler R, Blanco AC, dos Santos TF, Marya S. The effect of process parameters and tool geometry on thermal field development and weld formation in Frictio stir welding of the alloys AZ31 and AZ 61. In: Essential Readings in Magnesium Technology. Willey; 2014. pp. 529-533

[24] Myśliwiec P, Śliwa RE. Friction stir welding of thin sheets of magnesium alloy. Archives of Metallurgy and Materials. 2018;**63**(1):45-54

[25] Hadasik E, Kuc D. Deformation mode in AZ31 magnesium alloy during compression test. Acta Metallurgica Slovaca. 2010, s. 261-267;**2**:2

[26] Kuc D, Hadasik E, Niewielski G, Płachta A. Structure and plasticity of the AZ31 magnesium alloy after hot deformation. Journal of Achievements in Materials and Manufacturing Engineering. 2008;**27**(1):27-31

[27] Dziubińska A, Gontarz A, Dziubiński M, Barszcz M. The forming of magnesium alloy forgings for aircraft and automotive applications. Advances in Science and Technology Research Journal. 2016; **10**(31):158-168

[28] Śliwa R, Balawender T, Eugeniusz H, Dariusz K, Gontarz A, Korbel A, et al. Metal forming of lightweight magnesium alloys for aviation applications. Archives of Metallurgy and Materials. 2017;**62**:1559-1566

[29] Hadasik E. Badania plastyczności stopu magnezu AZ31. Rudy i Metale Nieżelazne. 2008;**R53**(12):790-794

[30] Hadasik E. Badania Plastyczności Metali. Gliwice: Wydawnictwo Politechniki Śląskiej; 2008

[31] Hadasik E, Kuc D, Bednarczyk I. Wpływ parametrów odkształcenia na gorąco na mikrostrukturę i właściwości plastyczne stopu magnezu AZ31. Hutnik-Wiadomości Hutnicze. 2017;**8**: 306

[32] Hadasik E, Kuc D. Plastic forming of magnesium alloys. Obróbka Plastyczna Metali. 2013;**XXIV**(2):133-146

[33] Al-Samman T, Li X, Ghush Chowdhur S. Orientation dependent slip and twinning during compression and tension of strongly textured magnesium AZ31 alloy. Materials Science and Engineering: A. 2010;**527**(15):3450-3463

[34] Gronostajski Z, Paterb Z, Madej L, Gontarz A, Lisiecki L, Lukaszek-Solek A, et al. Recent development trends in metal forming. Archives of Civil and Mechanical Engineering. 2019;**19**: 898-941

[35] Mineta T, Hasegawa K, Sato H. High strength and plastic deformability of Mg–Li–Al alloy with dual BCC phase produced by a combination of heat treatment and multi-directional forging in channel die. Materials Science and Engineering: A. 2020;**773**:31

[36] Dieringa H, Bohlen J, Hort N, Letzig D, Kainer KU. Advances in manufacturing processes for magnesium alloys. In: Mathaudhu SN, Luo AA, Neelameggham NR, Nyberg EA, Sillekens WH. Essential Readings in Magnesium Technology. 2014

[37] Kainer KU. Magnesium Alloys and Technology. Weinheeim: WI VCH Verlag GmbH & Co. KGaA; 2003

[38] Al-Samman T. Comparative study of the deformation behavior of hexagonal magnesium–lithium alloys and conventional magnesium AZ31 alloy. Acta Materialia. 2009;**57**: 2229-2242

[39] Palbam Metal Works, MagForming – FP6 project on Development of New Magnesium Forming Technologies for the Aeronautics Industry, Forming and SPF technologies development for introducing wrought magnesium applications in aerospace. Germany: AMTS, Palbam; 2006

[40] Bettles C, Barnett M. Advances in Wrought Magnesium Alloys, Woodhead Publishing Series in Metals and Surface Engineering; 2012. 10- Rolling. pp. 346-375

[41] Papenberg NP, Gneiger S, Weißensteiner I, Uggowitzer PJ, Pogatscher S. Mg-alloys for forging applications—A review. Materials. 2020;**13**:985. DOI: 10.3390/ma13040985

[42] Gontarz A, Pater Z, Drozdowski K. Hammer forging process of lever drop forging from AZ31 magnesium alloy. Metalurgija. 2013;**52, 3**:359-362

[43] Kuc D, Hadasik E, Gontarz A. Magnesium Forming Technologies. Mat. Konf. Wrought Magnesium Alloys, Resource - efficient Production and Applications Magnesium Alloy. Freiberg - Germany: Technische Universitat Bergakademie Freiberg; 2012. pp. 15-18

[44] Gontarz A. Kucie Matrycowe Stopów Magnezu. Lublin: Politechnika Lubelska; 2016

[45] ASM Metal Handbook, Volume 14A. Metalworking: Bulk Forming. In: Semiati SL, editor. ASM International. Materials Park, Ohio 44073-0002, 2005

[46] ASM Metal Handbook, Volume 14. Forming and forging. In: Davis JR, Semiatin L, American Society for Metals. ASM International; 1996

[47] Gontarz A, Drozdowski K, Michalczyk J, Wiewiórowska S, Pater Z, Tomczak J, et al. Forging of Mg-Al-Zn magnesium alloys on screw press and, forging Hammer. Materials. 2021;**14**:32

[48] Swistok J, Göken, Leitzig D, Kainer KU. Hydrostatic extrusion of commercial magnesium alloys and its influence on grain refinement and mechanical properties. Materials Science and Engineering A. 2006;**424**:223-229

[49] Śliwa RE, Pawłowska B, Balawender T. Metallic lightweight composite profiles for aviation obtained in extrusion process. In: AVT-264 Specialists Meeting on Design, Manufacturing and Application of Metallic Lightweight Material Components for Military Vehicles (NATO – Unclassified + Sweden +Australia and New Zealand), NATO Conference; April 2016; Tallin, Estonia

[50] Peel CJ, Gregson PJ. Design Requirements for Aerospace Structural Materials, High Performance Materials in Aerospace. Chapman & Hall; 1995

[51] Saha PK. Aerospace Manufacturing Processes. CRC Press; 2017

[52] Mihriban O. Pekguleryuz, Kainer K, Kaya A. Fundamentals of Magnesium Alloys Metallurgy. Oxford, Cambridge, Philadephia, New Delhi: Woodhead Publishing; 2013

[53] Bochniak W. Teoretyczne i praktyczne aspekty plastycznego kształtowania metali – Metoda KoBo. Kraków: Wydawnictwo AGH; 2009. p. 2009

[54] Bettles C, Barnett M. Advances in Wrought Magnesium Alloys, Fundamentals of Processing, Properties

and Applications, Woodhead Publishing Series in Metals and Surface Engineering. 2012

[55] Azushima A, Kopp R, Korhonen A, Micari F, Lahoti GD, Groche P, et al. Severe plastic deformation (SPD) processes for metals. CIRP Annales. 2008;**57**(2):716-735

[56] Cizek L, Rusz S, Hilser O, Śliwa R, Kuc D, Tański T, et al. Microstructure and propierities of serlected magnesium alloys prepared for SPD processing technology. Archives of Metallurgy and Materials. 2017;**62**(4):2365-2370

[57] Korbel A, Bochniak W. Method of plastic forming of materials. 1998. U.S. patent 5.737.959 European Patent 0.711.210, 2000

[58] Bochniak W, Marszowski K, Korbel A. Theoretical and practical aspects of the production of thin-walled tubes by the KOBO metod. Journal of Materials Processing Technology. 2005; **169**:44-53

[59] Balawender T, Zwolak M, Bąk Ł. Experimental analysis of mechanical characteristics of KoBo extrusion method. Archives of Metallurgy and Materials. 2020;**65**(2):615-619

[60] Pawłowska B, Śliwa RE. Recykling wiórów aluminium metodą KOBO. Obróbka Plastyczna Metali. 2017; **XXVIII, 4**:301-316

[61] Korbel A, Bochniak W, Śliwa RE, Ostachowski P, Łagoda M, Kusion Z, et al. Nisko-temperaturowa konsolidacja wiórów z trudnoodkształcalnych stopów aluminium. Obróbka Plastyczna Metali. 2016;**XXVII**(2):133-152

[62] Zwolak M, Śliwa RE, An effect geometry, parameters of surface structure and material properties of KOBO extrusion dies on plastic flow of metals to produce lightweight profiles for aviation. 11th Aluminium Two Thousand World Congress 9–13 APRIL 2019 TREVISO

[63] Zwolak M, Śliwa RE, Pawłowska B. Analysis of plastic flow of metallic materials with different billet structure under the conditions of the KOBO extrusion process. In: Conference Proceedings FORMING, Ladek Zdrój. 2018

[64] Zwolak M, Sliwa RE, Pawłowska B. Kształtowanie mikrostruktury wyciskanego metalu w różnych warunkach procesu KoBo. In: Materiały XIII Konferencji Naukowej nt.: Odkształcalność Metali i Stopów. Omis, Łańcut: Oficyna Wydawnicza PRz; 2019

[65] Thomas WM, Dolby RE. Friction Stir welding developments. In: David SA, Deb Roy T, Lippold JC, Smartt HB, Vitek JM, editors. 6th Int. Trends in Welding Research. Materials Park, Ohio, USA: ASM International; 2003. pp. 203-211

[66] Thomas WM, Nicholas ED, Needham JC, Murch MG, Temple-Smith P, Dawes CJ. Friction stir butt welding. International Patent Application no. PCT/GB92/02203. 1991

[67] Seidel T, Reynolds AP. Visualization of the material flow in AA2195 friction-stir welds using a marker insert technique. Metallurgical and Materials Transactions A. 2001;**32**(11):2879-2884. DOI: 10.1007/s11661-001-1038

[68] Mysliwiec P. Analiza efektu uplastycznienia cienkich blach ze stopów aluminium i magnezu w procesie zgrzewania tarciowego z przemieszaniem do zastosowania w konstrukcjach lotniczych i samochodowych. PhD Thesis, Rzeszów 2020

[69] Śliwa RE, Myśliwiec P, Ostrowski R, Bujny M. Possibilities of joining different metallic parts of structure using friction stir welding methods.

Procedia Manufacturing. 2019;**27**: 158-165

[70] Balawender T, Śliwa RE, Gałaczyński T. Friction stir welding of light alloys. 9th Int. Conf. AIRTEC "Supply on the wings". Frankfurt/Main; 2014

[71] Powell BR, Luo AA, Krajewski PE. Magnesium alloys for lightweight powertrains and automotive bodies. Chapter in book: Advanced Materials in Automotive Engineering. 2012. pp. 150-209

[72] Mordike BL, Kainer KU. Magnesium Alloys and their Applications. Wolfsburg: Werkstoff-Informationsgesellschaft; 1998

[73] Dobrzański LA, Bamberger M, George ET. Magnesium and Its Alloys: Technology and Applications. Teylor Francis Group; 2021

[74] Dorbane, Ayoub G, Mansoor B, Hamade RF, Kridli G, Shabadi R, et al. microstructural observations and tensile fracture behavior of FSW twin roll cast AZ31 Mg sheets. Materials Science & Engineering A. 2016;**649**:190-200 FSW

[75] Lacki P, Kucharczyk Z, Śliwa RE, Gałaczyński T. Effect of tool shape on temperature field in friction stir spot welding. Archives of Metallurgy and Materials. 2013;**58**:597-601. DOI: 10.2478/amm-2013-004

Section 2

Magnesium Alloys and Magnesium-Based Composites for Biomedical Applications

Chapter 4

Amorphous and Crystalline Magnesium Alloys for Biomedical Applications

Katarzyna Cesarz-Andraczke, Aneta Kania, Katarzyna Młynarek and Rafał Babilas

Abstract

Amorphous and crystalline magnesium alloys, developed for medical applications – especially implantology – present the characteristics of biocompatible magnesium alloys (Mg-Zn, Mg-Zn-Ca, Mg-Ca etc.). This chapter provides a brief description of the role of magnesium in the human body and the use of Mg in medicine. It presents the concept of using magnesium alloys in medicine (advantages and limitations) and the scope of their potential applications (orthopedic implantology, cardiac surgery etc.). The chapter shows classification of magnesium alloys as potential biomaterials, due to their structure (amorphous, crystalline) and alloying elements (rare earth elements, noble metals etc.). The mechanism and in vitro degradation behavior of magnesium alloys with amorphous and crystalline structures are described. The chapter also discusses the influence of alloying elements (rare earth elements, noble metals) on the in vitro degradation process. It also presents the methods of reducing the degradation rate of magnesium alloys by modifying their surface (application of protective layers).

Keywords: magnesium alloys, metallic glasses, resorbable implants, in vitro degradation behavior, protection coatings

1. Introduction

Magnesium is one of the most common elements in nature. It constitutes 2.7% of the earth's crust and can be found in the form of minerals, such as dolomite, magnesite or kainite [1]. The mechanical properties of pure magnesium are poor, therefore alloying additives are introduced to improve them. Magnesium alloys, in which the major additive is aluminum – typically 6–10% – are the most widely used industrial alloys of magnesium. Zinc or manganese [2, 3] are added to improve corrosion resistance of magnesium alloys. Initially, magnesium alloys were produced mainly for military purposes. Due to the high specific strength and vibration damping capacity, magnesium alloys are mainly used in the automotive industry [3, 4].

Currently, research is carried out mainly on new groups of magnesium alloys, such as Mg-Ca, Mg-Zn and Mg-Zn-Ca, which have not been produced on an industrial scale so far. Studies are carried out on the use of these alloy groups as materials for implants, especially orthopedic ones. Injuries of the osteoarticular system, as well as diseases of the musculoskeletal system, including the continuous increase in the incidence of bone cancer, are the main and the most common threat to the health of modern society. 2,710,000 cases of orthopedic fractures were noted in Poland in 2010. Due to the aging of the population, it is predicted that, in 2025, their number will reach 3,239,564, and 10 years later – over 4 million. In 2017, 85,488 joint arthroplasties (partial or total) were performed in Poland, including 56,688 of the hip and 27,653 of the knee. Surgical joining of broken bones through their correct connection and immobilization is performed with the use of bone plates, wires, clamps and/or screws. The use of these elements results in bone union and obtaining the correct bone structure, which, in turn, allows the patient to move, and thus return to basic life activity [5–8].

In medical practice, both long-term implants (e.g. joint prostheses) and short-term implants (e.g. plates, bone screws), used to stabilize broken bones, are produced from titanium alloys, cobalt alloys or stainless steel. Implants made of those are classified as neutral, i.e. neutral to the body, as long as protective layers (usually oxide) remain on their surface. Unfortunately, after some time, these layers become corroded and damaged, and implant components – which are usually biologically incompatible (toxic to the body) – pass into the human body, and thus are a threat to health and life. Resorbable biomaterials are an alternative to the metal alloys used so far for short-term orthopedic implants. The resorbable biomaterials used in medical practice so far include oxide glasses (composed of Na, K, Mg, Ca, Si and P oxides), ceramics based on calcium phosphates, e.g. hydroxyapatite [$Ca_{10}(PO_4)_6(OH)_2$] and polymers, such as polylactide, polyglycolide or copolymers of these materials. Unfortunately, their use in orthopedic implantology is limited. They are mainly used as fillers for bone defects, elements of dentures or coatings for medical implants [9–11]. This is due to poor mechanical properties of resorbable biomaterials. Mechanical strength of the resorbable materials used in medicine is 30–100 MPa [10, 12]. Consequently, mechanical properties of resorbable polymers and ceramics are a barrier to their use as biomaterial for implants, such as short-term orthopedic implants. Accordingly, resorbable metallic biomaterials that can be used for orthopedic implants are necessary. Resorbable metallic biomaterials are an alternative to the metal alloys previously used for short-term orthopedic implants. Magnesium alloys are appropriate materials for resorbable metallic biomaterials.

This chapter presents the role of magnesium in the human body and its use in medicine. It presents the concept and potential applications of magnesium alloys in medicine, as well as classification of magnesium alloys as potential biomaterials due to the structure (amorphous, crystalline) and alloying elements (rare earth elements, noble metals etc.). The chapter also describes mechanisms and degradation behavior (in vitro) of magnesium alloys due to their structure. The impact of alloy additives (rare earth elements, noble metals) and protective coatings on the degradation process of magnesium alloys for biomedical applications in in vitro conditions has also been assessed.

2. The role of magnesium in the human body and its application in medicine

Magnesium is called an element of life, because it participates in many processes of the human body. It is necessary to maintain proper homeostasis, i.e. the proper functioning of the human body. It is estimated that there are approximately 22–26 g of magnesium in the human body [13]. It should be mentioned, that both the value and the range of the concentration of an element in the human body depends on the age, sex, absorption of the elements or even diet. The World Health Organization (WHO) has issued standards defining the daily demand for the element [14, 15].

Similarly, demand for magnesium is different for a certain age. **Table 1** shows the daily magnesium requirement depending on age.

Magnesium is distributed in the body (in the skeletal system: approx. 60%; in skeletal muscles: approx. 20% and other soft tissues: approx. 19%). Magnesium in acidic form, absorbed from food, in about 30%, can be found predominantly in the small intestine. The daily recommended dose of magnesium depends on the age, gender and current condition of the body. It has been proven, that the average dose for an adult human is about 300–400 mg. Approximately 30% of magnesium is absorbed from the gastrointestinal tract. Absorption of this element is influenced, among others, by the amount of consumed protein, fiber and phosphates. The normal blood magnesium level of a healthy person is 0.75–0.95 mmol/dm^3, and its homeostasis is maintained by the kidneys.

Magnesium has a lot of functions in the human body. For example, it:

- regulates the activity of about 300 enzymes involved in metabolic changes,
- is necessary for proper bone mineralization. It has been confirmed that magnesium deficiency disturbs bone mineralization processes, increasing the incidence of postmenopausal osteoporosis [17],
- is involved in nerve conduction and muscle contractility. It is likely that magnesium could be used to treat affective disorders and depression. A positive effect of magnesium on depression symptoms has been demonstrated in patients with low levels of magnesium in erythrocytes [18],
- plays a vital role in most hormonal responses. Magnesium has been shown to influence insulin synthesis, catecholamine storage and parathyroid hormone release,
- participates in the regulation of blood pressure,
- regulates muscle tension,
- regulates the thyroid gland and widens the airways, supporting the treatment of asthma and bronchitis.

Magnesium has been used in treatment of various diseases. **Figure 1** presents the main uses of magnesium for treatment of diseases.

Symptoms of magnesium deficiency influence every system in the human body. The most common symptoms are not very specific – they include fatigue, poor concentration and memory, as well as increased susceptibility to stress. Excessive loss of magnesium can be caused by serious diseases of the gastrointestinal tract (e.g. fistulas, pancreatitis), urinary tract disorders, endocrine disorders (primary

DEMAND FOR MAGNESIUM [mg/24 h]					
Infants	kids < 6 years old	kids 6–9 years old	youth 10–18 years old	adults 19–60 years old	adults < 60 years old
40–60	**80–120**	**170**	**270–400**	**280–350**	**280–350**

Table 1.
Daily recommended dose of magnesium for kids, youth and adults [16].

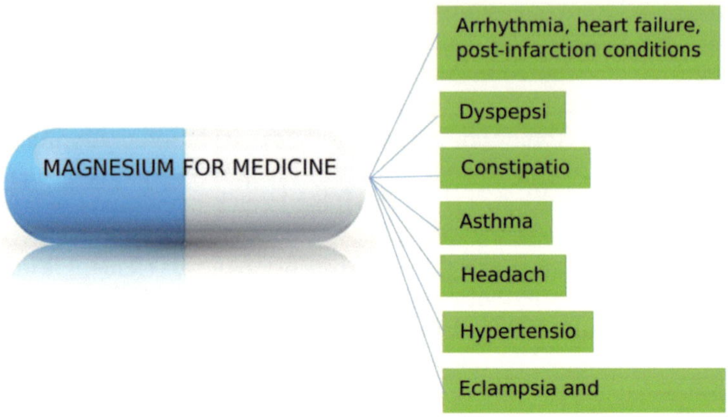

Figure 1.
The most important uses of magnesium in treatment of diseases [19].

hyperparathyroidism, intensive insulin therapy), D3 hypervitaminosis, use of immunosuppressants, increased sympathetic nervous system tension or alcoholism.

3. The concept and potential applications of magnesium alloys in medicine (advantages and limitations)

The progress of both the medicine and materials engineering results in an intensification of research works on new biomaterials. Nowadays, magnesium alloys are considered potential resorbable metallic biomaterials. Furthermore, it is assumed that a magnesium alloy as a resorbable biomaterial should gradually degrade in the human body until the bone fuses. Degradation products of a resorbable implant would be processed, absorbed or excreted from the patient's tissues and body fluids. The use of implants designed according to this concept does not require re-operation and it allows the foreign object (implant) to stay in the human body. Apart from good mechanical properties and biocompatibility, magnesium has a number of other advantages, such as [20, 21]:

- good strength-to-weight ratio. Pure magnesium has 158 kNm/kg; however, its alloys can reach up to 490 kNm/kg. This is twice as much as the most commonly used titanium alloys (260 kNm/kg), therefore, less material is needed to obtain similar mechanical properties.

- ease of processing magnesium and creating complex shapes, which is extremely important in medical applications, because every person is different and, therefore, it is possible to design a custom-made implant for a specific patient.

- safe degradation – titanium, stainless steel and Co-Cr alloys do not ensure safe degradation. All surgically implanted alloys are subject to electrochemical degradation, as they are in a corrosive environment. Additionally, they are subject to significant wear. Implant particles can be released into the surrounding tissue, causing discomfort and potential health hazards. Magnesium and its alloys can minimize these problems during the degradation process. It is possible that, after a controlled period of time, the implant completely degrades in the human body.

The main problems and research limitations of this concept are as follows:

- production of implants with good mechanical properties, which guarantee the appropriate time (allowing bones to fuse) of the implant's activity in human bodily fluids,

- high degradation of an implant with very intense release of hydrogen, which is harmful to the body. In addition, there is possible exceeding the daily demand for the element (also biocompatible), introducing into the body. Metal alloys used for resorbable biomaterials should only include elements that are already present in the human body in high concentration and are macro- or microelements.

The major problem concerning all metal biomaterials consists in the adjustment of their mechanical properties to those of the reconstructed tissues. The density of steel is approx. 4 times higher than the density of bone tissue. Steel has several times higher yield point and tensile strength, higher elongation and about 10 times higher Young's modulus. The differences in the mechanical properties of materials and those of the tissues they replace result in inappropriate loading of the tissues surrounding the implant, causing pain and discomfort in patients [22].

The density of magnesium alloys is similar to the density of the human bones, therefore, there is no possibility of stress shielding, as in the case of the previously used implant materials, based on stainless steel and titanium. Stress shielding is a process, in which the bone mass and density decrease near the implant, because it transfers the loads. The density of magnesium alloys is three times lower than that of titanium alloys and five times lower than that of stainless steel and Co-Cr alloys. The modulus of elasticity and fracture resistance are much lower than for the biomaterials used so far [23, 24]. In the 19th and 20th centuries, magnesium alloys were used in medicine (**Table 2**). They were used in the form of scaffolds and to improve healing of wounds or organs. In addition, magnesium and its alloys have

Author	Application date	Mg/Mg alloys	Application
Huse	1878	Pure magnesium	Stitches
Payr	1892–1905	Pure magnesium	Nerve's linkers
Hopfner	1903	Pure magnesium	Vessels' linkers
Lambotte	1906–1932	Pure magnesium	Bone screws and plates
Lespinasse	1910	Pure magnesium	Bone plates
Groves	1913	Pure magnesium	Bone pits
Andrews	1917	Mg-Al/Zn	Bone wires
Seelig	1924	Pure magnesium	Bone wires
Verbrugge	1933–1937	Mg-Al6-Zn3-Mn	Bone screws and plates
McBride	1938	Mg-Mn	Bone wires and plates
Maier	1940	Pure magnesium	Stitches
Stone	1951	Mg-Al (2 wt.%)	Bone wires
Wexler	1980	Mg-Al (2 wt.%)	Bone wires
Hussl	1981	Pure magnesium	Vascular wires

Table 2.
Applications of magnesium and its alloys in medicine [25].

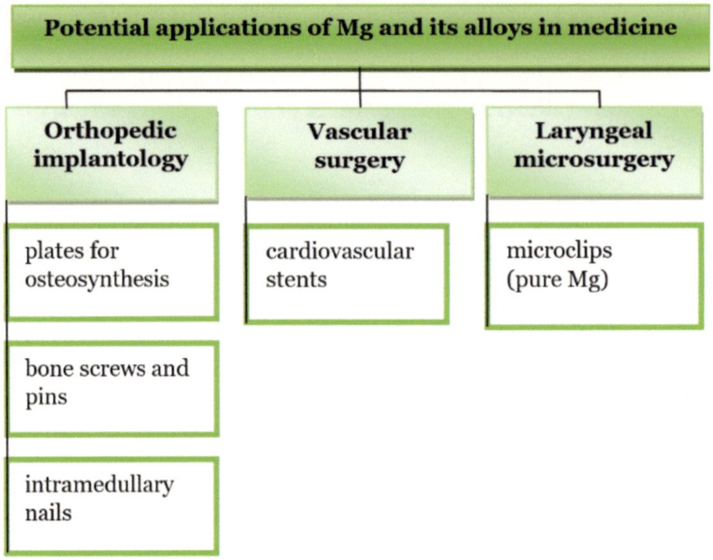

Figure 2.
Potential applications of Mg and its alloys in implantology [17–22].

been used in orthopedic surgery for such elements as screws, plates, fasteners or as stents in the cardiovascular system.

At present, magnesium alloys are mainly considered as potential materials for applications in orthopedic implantology, vascular surgery and laryngeal microsurgery [26–30]. As regards orthopedic implantology, Mg alloys are used as compression screws. MAGNEZIX is the trade name of biodegradable orthopedic screws for human osteosynthesis application [31]. In the literature, there are some reports [32–34] on resorbable stents made of magnesium alloys (their trade name is Lekton Magic, produced by Biotronik company). This material is composed of zirconium (< 5 wt.%), yttrium (< 5 wt.%) and rare earth elements (< 5 wt.%). The stents degrade in a living body with time, but their location can still be identified. Finally, the stent material completely degrades and the space around it is filled with a calcium-apatite complex with an admixture of phosphate elements. The stents were implanted in 20 people and good flow in the implanted blood vessel was achieved after one month. In 2013, Biotronik, a German company, has obtained the CE mark for biodegradable coronary stents made of Mg alloy. It was the leader in development of biodegradable metal coronary stents. The areas of potential applications of magnesium and its alloys in implantology are presented in **Figure 2**.

4. Classification of magnesium alloys considered as potential biomaterials due to their structure (amorphous or crystalline) and alloying elements (rare earth elements, noble metals etc.)

In the context of resorbable orthopedic implants, research was initially carried out on technical magnesium alloys, for example AZ31, AZ91, WE43, LAE442. Unfortunately, magnesium alloys containing aluminum (AZ31) and heavy metals have been excluded as biomaterials because these additives have a toxic effect on the human body. The research was limited to alloys containing biocompatible elements and/or small amounts of rare earth elements, that are tolerated by the human body in appropriate concentrations [15].

As regards magnesium alloys for resorbable implants with a crystalline structure, the following groups of alloys has been examined: Mg-Ca, Mg-Zn, Mg-Zn-Ca, Mg-Mn, Mg-Si, Mg-Zr, Mg-Zn-Zr, Mg-Zn-Y, Mg-Zn-Zr-Y, Mg-Zn-Mn.

Rare earth elements (REE) are added in order to improve mechanical properties and creep resistance at elevated temperatures [35, 36]. Gadolin (Gd) and yttrium (Y) increase the strength properties during precipitation hardening. Neodymium (Nd) improves tensile strength at ambient and elevated temperatures. Yttrium and strontium (Sr) reduce the texture, and thus anisotropy, in rolled and extruded semi-finished products [35].

Magnesium alloys with the addition of rare earth elements, such as Mg-Y, Mg-Gd [37] and Mg-Nd, have been designed for use as biomaterials. ZW21 and WZ21 alloys (with the addition of Y and Zn) show promising mechanical and corrosion properties. For example, they are ductile (up to 20% elongation) and their tensile strength $\approx$ 270 MPa. Alloys such as AE21 and WE43 are used for stents [38, 39].

Noble metals as an additive to magnesium and calcium alloys have been studied mainly by the authors of this chapter [12, 40]. There is no information in the literature on the influence of Au and Pt addition on the degradation rate and mechanical properties of magnesium alloys. There are several sources for adding silver to magnesium alloys [41, 42].

The mechanical and corrosion properties of the alloy can be regulated by the structural and chemical composition of the alloy. Compared to their crystalline counterparts, magnesium-based metallic glasses may be more resistant to corrosion, due to their single-phase structure, which may result in a more uniform alloy corrosion. An example confirming the higher corrosion resistance of the amorphous material in physical fluid compared to the crystalline material with the same chemical composition is shown in **Figure 3**.

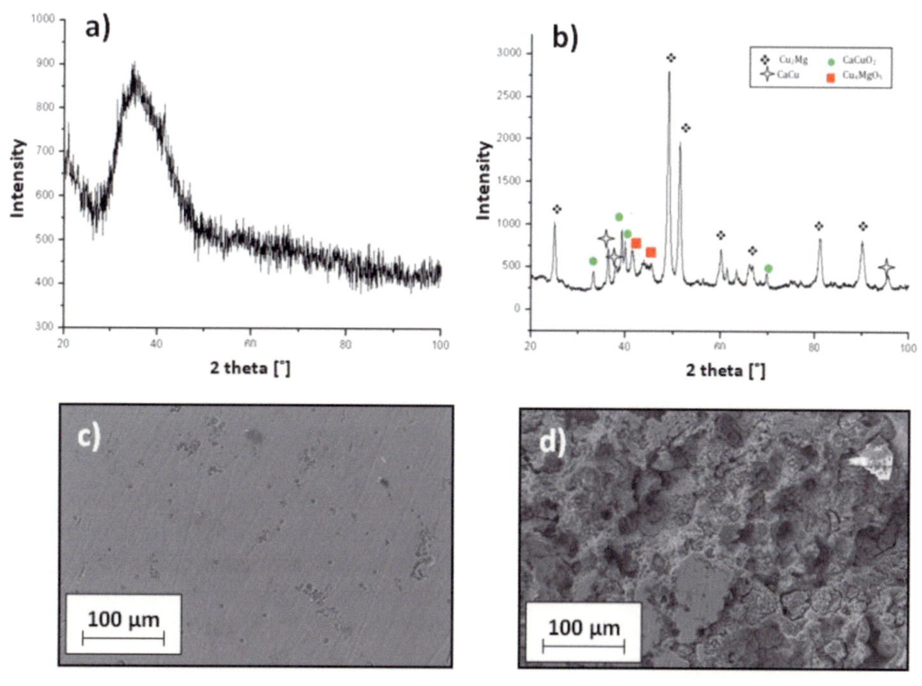

Figure 3.
Results of structural tests of Mg36.6Cu36.2Ca27.2 alloy with amorphous structure (a) and crystalline structure (b) and surface images after 1.5 h of immersion in physiological fluid with amorphous structure (c) and crystalline structure (d) [43].

Magnesium alloys with amorphous structure, such as bulk metallic glasses (e.g. rods, plates) in the following phase systems: Mg-Cu-Y (-Ag, -Pd, -Gd), Mg-Ni-Y (-Nd), Mg-Cu-Gd (-Zn, -Y), Mg-Zn-Ca were obtained. In addition, studies are also carried out on Mg-based metallic glass without rare earth elements. The Laws [44] obtained bulk metallic glasses based on Mg-Cu-Ca, Mg-Ag-Ca, Mg-Cu-Ag-Ca alloy systems. However, for applications in implantology, magnesium alloys should have a biocompatible chemical composition. Therefore, the group of alloys based on the Mg-Zn-Ca phase system is most frequently considered as a new biomaterial for resorbable orthopedic implants [45]. In 2005, Gu et al. were the first to obtain bulk metallic glass in the Mg-Ca-Zn system, which was characterized by good strength properties and high glass transition capacity [46].

In the process of designing new degradable biomaterials, elements with potential toxicological problems should be omitted whenever possible and, if they are absolutely necessary, they should be reduced to the minimum. Calcium and zinc are essential elements in the human body; therefore, these elements should be the first choice for alloying additives in biomedical magnesium alloys. The concentration of calcium should not exceed 2 wt.%, and zinc – 6 wt.%, due to the corrosive properties of these magnesium alloys [47].

The most commonly used chemical elements for magnesium alloys are: Zn, Zr, Ca, Sr., Yb, Al, Li, Mn and rare earth elements (REEs) (Ce, Er, La, Gd, Nd, Y). The following are the additions, the influence of which on the properties of magnesium alloys is described in detail:

- addition of zinc (< 5 wt.%) reduces the harmful influence of iron and nickel impurities, increases corrosion resistance [48],
- addition of zirconium (< 2 wt.%) increases corrosion resistance [48],
- addition of strontium (< 2 wt.%) improves corrosion properties and affects the strength of the alloy, which is similar to the natural bone [48, 49]. Optimal content of Zr and Sr. in Mg-based alloys increases surface energy and the ability to simulate contact osteogenesis. Mg-Zr-Sr alloys (2 wt.% Sr) display the best osseointegration and complete biodegradation [50],
- addition of ytterbium (at the level of 2 wt.%) improves bending plasticity, corrosion properties and biocompatibility [49],
- addition of calcium (> 1 wt.%) in pure Mg reduces corrosion resistance [48]. Calcium in magnesium alloys, without the addition of strontium, reduces surface energy and bone induction [50],
- addition of yttrium (> 2 wt.%) decreases corrosion resistance in Mg-Y alloys [48].

Noble metals, such as gold and silver, were used as alloying additives in pure magnesium to increase its ductility. However, the alloys had low tensile strength [51]. Another source mentions that the addition of silver, as a substitute for calcium, improves the corrosion properties, strength and has an anti-bacterial effect [49].

Figure 4 shows variation in the open-circuit potential with time and polarization curves for pure Mg and $Mg_{65}Zn_{20.1}Ca_{1.7}Yb_{13}Sr_{0.2}$ alloy in Ringer solution at 37°C.

In the OCP plot (**Figure 4a**), various levels of recorded curves are visible, which results from differences in chemical compositions. The steady-state for

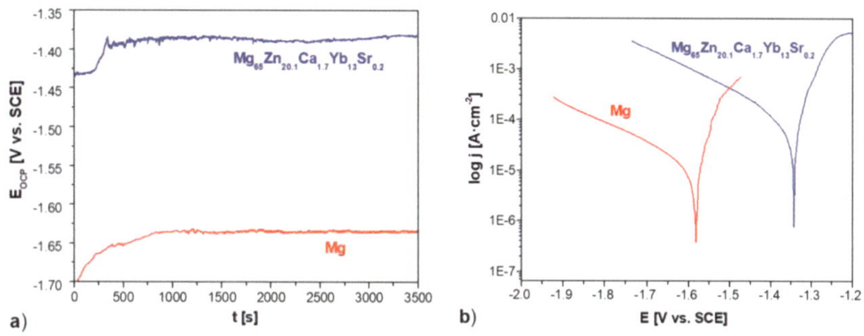

Figure 4.
Variation of the open-circuit potential with time (a) and polarization curves (b) for pure Mg and $Mg_{65}Zn_{20.1}Ca_{1.7}Yb_{13}Sr_{0.2}$ alloy in Ringer solution at 37°C.

pure magnesium is in the range of approx. -1.65 V, while for $Mg_{65}Zn_{20.1}Ca_{1.7}Yb_{13}Sr_{0.2}$ alloy – slightly above −1.4 V. This shift towards positive values indicates favorable behavior of samples with alloy additions. Potentiodynamic measurements (**Figure 4b**) also show the differences between the studied alloys. The values of corrosion current density were slightly higher for magnesium alloy, as compared to pure Mg. However, significant differences in E_{corr} by approx. 0.25 V are observed, which indicates that it is recommended to use alloying elements to improve corrosion resistance. E. Mostaed et al. [52] showed similar results regarding the differences in electrochemical tests between pure magnesium and the ZK60 alloy.

Designing of magnesium alloys as biomedical materials is a great challenge, due to rapid degradation of Mg in the environment of bodily fluids and insufficient implant-bone connection in orthopedic applications [50]. These disadvantages can be limited due to an appropriate selection of alloying additions. The purpose of optimal chemical composition of a new class of Mg-based biodegradable materials is to obtain optimal strength, ductility, resistance to fatigue and corrosion by modifying the structure and phase distribution [50, 53]. Currently, efforts are being made to select alloying additions that would promote osseointegration, understood as the fusion of the implant with the newly formed bone tissue and biodegradation without adverse effects on the functioning of body organs [50].

5. Mechanisms and in vitro degradation behavior of amorphous and crystalline magnesium alloys

In the case of biomedical engineering, corrosion is the main factor determining the usefulness of implant materials. The tendency of biomaterials to corrode in the human body is, in fact, closely connected to their biocompatibility. Before placing in the human body, the material must be examined for the effects on the body and its properties. Ensuring such experimental conditions is difficult, as it is difficult to recreate the environment of the human tissues. Many parameters related to the production of magnesium-based materials and test parameters have an impact on the degradation results (**Figure 5**).

The alloying elements and the processing parameters of Mg have a strong impact on its degradation properties (microstructure of the material described by the grain size, impurity content, type of phases etc.). Calcium as alloying element to Mg alloy is an extremely reactive metal and spontaneously reacts with water generating hydrogen [55].

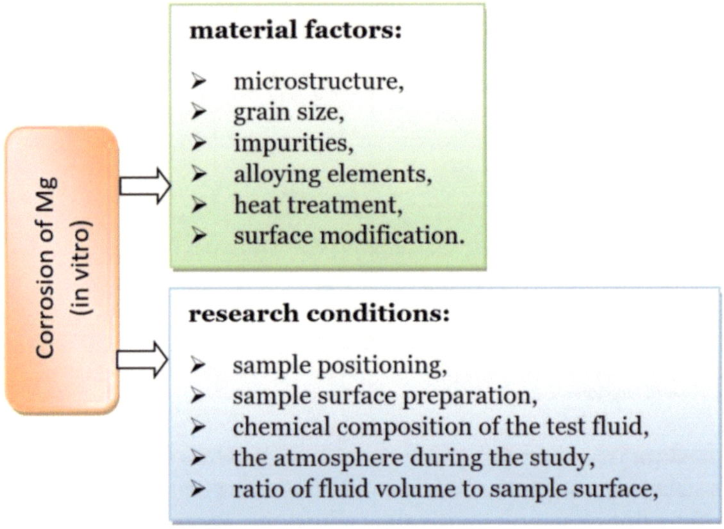

Figure 5.
Parameters influencing the course of magnesium alloy corrosion process (in vitro) divided into subgroups: Research conditions and material factors [54].

Moreover, the research methods and conditions can significantly change the corrosion rate, as well as the formation and composition of the degradation layer, and thus determine the corrosion type [56].

Living microorganisms play an important role in the process of implant degradation. Such metabolic activity may directly or indirectly reduce the quality of the implant due to the corrosion process. Cells can act as an electrolyte on the metal surface, thus changing the corrosion resistance of the implant surface or even its composition [6].

Corrosion of magnesium in an aqueous environment occurs as a result of an electrochemical reaction with water, resulting in the formation of magnesium hydroxide, $Mg(OH)_2$ and hydrogen gas, according to reactions (1–3) [54]:

$$Anodic\ reaction: Mg \rightarrow Mg^{2+} + 2e \tag{1}$$

$$Cathodic\ reaction:: 2H_2O + 2e^- \rightarrow H_2 + 2OH^- \tag{2}$$

$$Overall\ reaction\left(summary\ reaction\right): Mg + 2H_2O \rightarrow Mg\left(OH\right)_2 + H_2 \tag{3}$$

In the initial phase of immersion, the surface of the material is exposed to the electrolyte and the anodic and cathodic reactions begin. Magnesium grains act as an anode and the cathodic reaction takes place in noble regions of alloy, which are grain boundaries, phase separation and precipitation. This leads to the exchange of electrons between the two regions, wherein the magnesium is degraded at the same rate, at which hydrogen is generated as a gas (H_2). The cathodic reaction increases the pH by releasing H_2 gas, while hydrolysis lowers it [54, 57].

When the concentration of Mg^{2+} and the increase of pH reach the solubility limit, magnesium hydroxide $Mg(OH)_2$ is precipitated on the surface of alloy Mg [1]. In an environment, such as body fluids, where the concentration of chloride ions is greater than 30 mmol/dm^3, the hydroxide formed on the surface of the magnesium alloy converts to highly soluble magnesium chloride. This reduces the level of protection of the surface layer by increasing its activity [58]. The formation of soluble magnesium chloride is described by the reaction (4):

$$Mg(OH)_2 + 2Cl^- \rightarrow MgCl_2 + 2OH^- \quad (4)$$

This process accelerates the degradation of the material and increases the pH of the environment [6]. The presence of Cl^- ions initiates pitting corrosion.

It should be mentioned that the structure of magnesium alloys mainly affects the course and rate of the degradation process. The analysis of corrosion tests results and studies of degradation products on the surface of the amorphous $Mg_{64}Zn_{32}Ca_4$ alloy allow to distinguish and link the probable stages of the degradation process for the tested alloy in selected micro-areas, which include [59]:

- transformation of the oxide layer into hydroxide,
- penetration of the hydroxide layer by chlorides,
- release of metallic ions and their transfer to solution,
- hydrogen evolution,
- creating a protective layer.

It should be noted, that the specified steps are not consecutive, but may occur simultaneously during the immersion of the amorphous $Mg_{64}Zn_{32}Ca_4$ alloy. Therefore, the degradation of the amorphous $Mg_{64}Zn_{32}Ca_4$ alloy can be considered as a total result of the following processes: the release of alloy components and the formation of protective layers. When the sample is immersed in a chloride solution, degradation occurs directly at the surface, due to the rapid release of active Mg and Ca. On the other hand, this results in enrichment of the sample's surface with less active zinc. With the progress of degradation, zinc is oxidized and accumulates in the vicinity of disturbed chlorides, and therefore protects against further progression of degradation [60, 61]. However, the protective layer is not dense enough to completely prevent degradation. Chlorine ions damage the zinc oxide layer. Damage to the protective layer facilitates the transition to the Ca and Mg ion solution. These mechanisms are repeated until the amorphous magnesium alloy has degraded completely [62].

6. Methods of reducing the degradation rate of magnesium alloys by modifying their surface (application of protective layers)

High corrosion rate of magnesium-based alloys in tissue environment may be limited, in addition to modifying the chemical composition, by surface treatment technologies. The degradation process can be controlled by way of coating the surface or changing its structure [53, 60, 61]. There are two methods of coating: conversion and deposition processes. Conversion coatings are the product of complex interaction of metal dissolution and precipitation, usually during treatment in aqueous solutions, while deposition treatments consist of metallic, inorganic and organic coatings [53, 62, 63]. Modifications in the surface of magnesium alloys by mechanical treatment are also used [62]. The classification of the coating technology for magnesium alloys is shown in **Figure 6**.

Homogeneity of the corrosion process is an important aspect that determines the degradation rate and the physical condition of the implant at a specific treatment

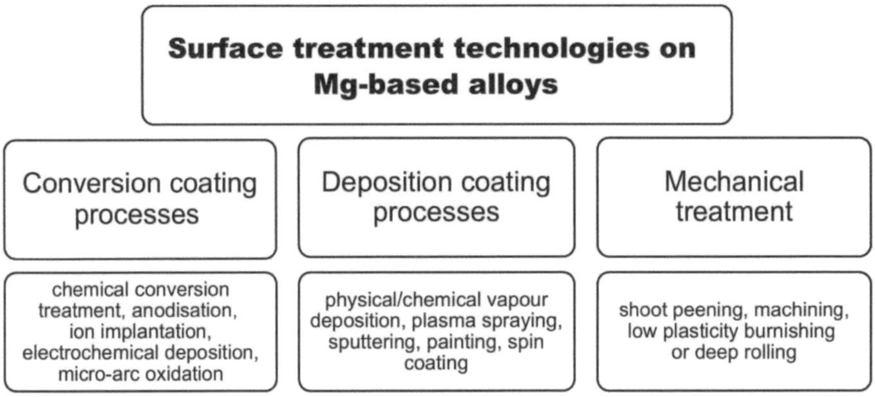

Figure 6.
General classification of surface treatment technologies applied on magnesium alloys [62, 63].

stage [63]. Magnesium alloy coatings often have pores and cracks. Corrosion, which begins in these areas, leads to uneven rate of corrosion, accelerates destruction of the coating and premature degradation of the implant [63, 64]. Therefore, it is important to minimize the porosity of the coating by adjusting the parameters of the application process or the appropriate preparation of the substrate's surface [62, 65]. In addition, protective coatings on biodegradable magnesium alloys should be adapted to specific applications – e.g. vascular stents have different surface requirements than orthopedic implants, where osseointegration with newly formed bone is important [62, 63]. Selected technologies of forming coatings on magnesium alloys are discussed below with regard to their advantages and disadvantages in terms of use, with an aim to reduce the corrosion rate:

- Chemical conversion treatment – coatings based on $Mg(OH)_2$ and fluorine are the most commonly used coatings developed with this technique, increasing corrosion resistance, while remaining non-toxic for the surrounding tissues [63, 66]. Their main advantage is good adhesion [67]. Chemical conversion treatment is still considered an economically viable technique [63, 66].

- Micro-arc oxidation (MAO) – is considered the most cost-effective technology in the production of protective coatings against corrosion on Mg-based alloys. Surface pores and cracks, which accelerate the corrosion rate, are a significant disadvantage of this technology [60, 63].

- Electrochemical deposition (ED) – is one of the most widely used methods. It has the ability to create homogeneous, dense protective layers, preventing further corrosion of the magnesium alloy substrate. Its great advantages, from technological point of view, are reproducibility and low temperatures of deposition [63, 66, 67].

- Anodization – the quality of the coating obtained in this technique is strongly dependent on the parameters of the process: electrolytic composition, applied constant voltage or current, quality of the alloy surface and concentration of the alloying elements. The obtained 5–200 μm thick oxide layer creates functional corrosion protection with excellent adhesion [63, 67]. The main disadvantage of anodization is low wear resistance [67]. The technique makes it possible to obtain nano-tubular porous layers. However, this kind of surface structure is not suitable for some applications, e.g. stents [62, 63].

- Ion Implantation – thin layers, resulting from the process, do not provide the required corrosion resistance. Despite many advantages of this technique, like modification of physical, chemical and electrical properties, ion implantation is expensive and is not suitable for complex geometries of implant components [62, 63, 67].

- Physical/Chemical Vapor Deposition (PVD/CVD) – known, widely used technologies [63]. PVD enables the formation of hard coatings, resistant to tribological wear, but from the perspective of corrosion resistance, the layers are too thin and have pores. CVD is energy-efficient and does not manifest toxicity, but has the disadvantages of complicated layer growth and requires high temperatures [67].

In choice of the coating technology, one should always take into account preservation of the alloy's biocompatibility, due to the potential toxicity of the elements introduced in coating treatments. In addition to coating, an alternative solution is mechanical treatment (shot peening, machining, burnishing, deep rolling), which solves the toxicity problem. The literature also describes hybrid techniques, which combine mechanical treatment with coating, as a promising solution for controlling the corrosion rate and mechanical properties at individual stages of treatment [63]. Biomimetic coatings are also noteworthy, as they are of biological origin and ensure excellent biocompatibility, but further work is required to improve their low adhesion to the substrate alloy [63, 67].

The types of protection coatings used to delay/reduce the degradation rate of magnesium alloys are shown in **Figure 7**. Besides phosphate and fluoride, most of the proposed ceramic coatings are non-resorbable. In the case of resorbable materials, considered polymeric coatings include PLA, PLGA and copolymers. Composite coatings increasing the corrosion resistance of magnesium alloys, tested by several researchers, include the following types of coatings: ceramic-metallic and ceramic-polymer.

As part of the authors' own research, tests of phosphate coatings on magnesium alloys were carried out. In this work [78], the chemical method was used for Ca-P coatings preparation. NaOH and $ZnSO_4$ as accelerators were added to phosphatizing baths, with an aim to form a dense and uniform protective phosphate coating. It

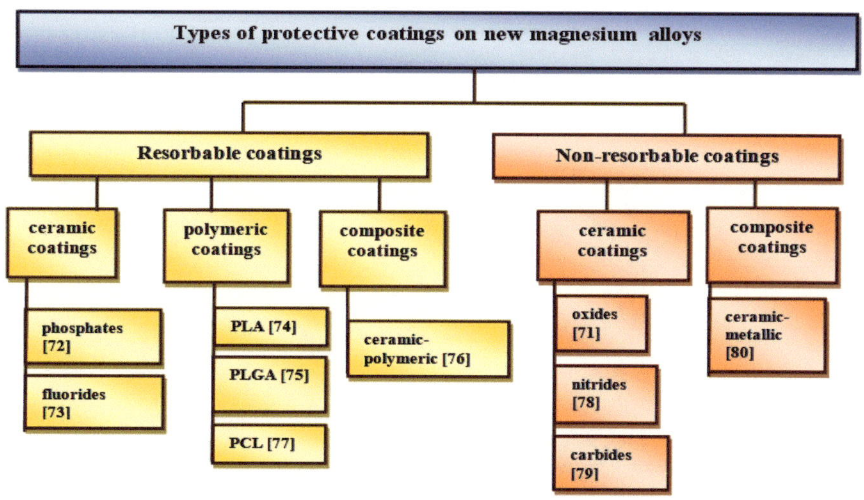

Figure 7.
Types of protective coatings used to delay/reduce the degradation rate of magnesium alloys [68–77].

should be noted, that NaOH and $ZnSO_4$ are used to improve corrosion resistance of Mg alloys. The results of microscopic observations and phase identification of the obtained phosphate coatings (with the use of chemical composition of the phosphating bath) are shown in **Figure 8**.

XRD results indicate that obtained protection layers included dicalcium phosphate dihydrate ($CaHPO_4{\cdot}2H_2O$). Both NaOH and $ZnSO_4$ formed the morphology of the produced layers. The coating obtained by immersion in a phosphatizing bath with $ZnSO_4$ addition (ZnAM50 sample) consisted of petals. The coating obtained by immersion in a bath with NaOH addition (NaAM50 sample) showed plate-like morphology.

The degradation tests of magnesium alloys with Ca-P layers were also performed (**Figure 9a** and **b**) in Ringer's solution at 37°C. The results of electrochemical tests indicated that coated samples have more positive value of E_{corr} than non-coated AM50 sample (**Figure 9a**). In addition, the cathodic part of potentiodynamic curve

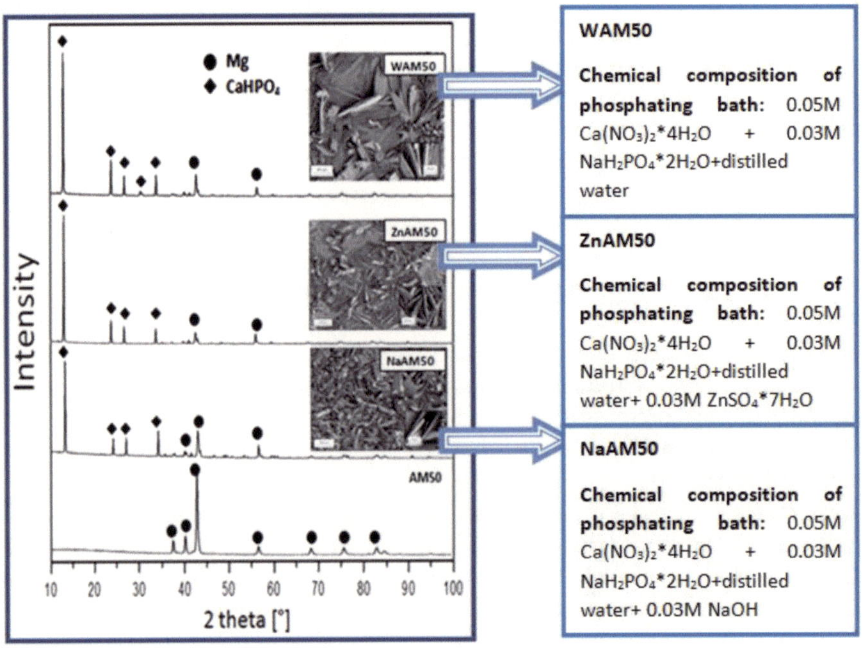

Figure 8.
X-ray diffraction patterns and SEM images of Ca-P coatings on Mg alloy [78].

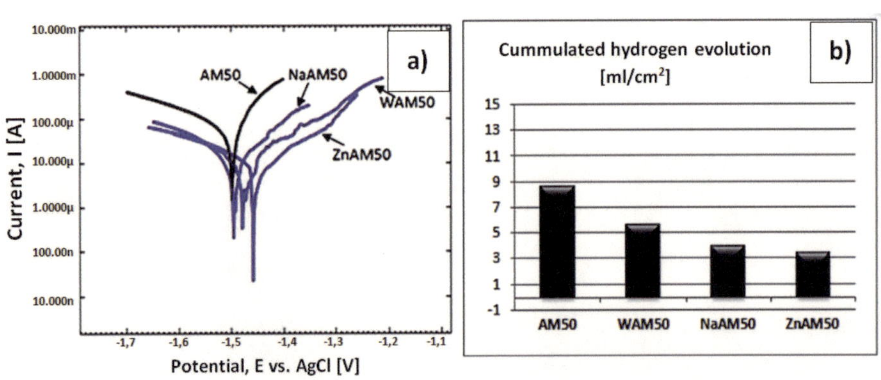

Figure 9.
Results of degradation tests of Mg alloys with calcium phosphate coatings in Ringer's solution at 37°C: (a) polarization curves, (b) hydrogen evolution [78].

determined for the coated samples is located in a low current range, which indicated a low cathodic activity. It corresponds with the immersion tests results (**Figure 9b**). The volume of evolved hydrogen (hydrogen is a result of cathodic reactions) in an uncoated sample was higher than its level in coated samples.

The degradation rate of ceramic material determines the occurrence of defects, cracks and flaws in technology. Defected ceramics can be destroyed in contact with water. Inclusions of other phases are equally disadvantageous to ceramic materials, that lead to their degradation. In contact with water, these inclusions accelerate aging and increase volume. These processes also have a direct impact on deterioration of the mechanical properties of ceramic materials [79].

7. Conclusions

Magnesium is a very important macroelement for the human body and serves a lot of functions. Its deficiency can cause many disorders and health ailments. Accordingly, magnesium is widely used in therapy, primarily for the treatment of heart disease, cardiovascular system or respiratory system. This became a premise for the use of magnesium and its alloys in medicine as a potential biomaterial for medical implants. The concept of using magnesium alloys as resorbable medical implants assumes that it will be non-toxic to the human body. The alloy components will also be elements present in the human body. The resorbable biomaterial of a magnesium alloy would be an alternative to the previously used implants, mainly orthopedic ones. Unfortunately, the high degradation rate of the magnesium alloy and the release of hydrogen gas in the environment of physiological fluids limit the use of these alloys as a biomaterial. Therefore, the research community continues to test different types of surface treatment for magnesium alloys, to protect it from rapid degradation. Taking into account the results of the global research and the authors' own research, this seems to be the right way to obtain a resorbable biomaterial of magnesium alloy.

Conflict of interest

The authors declare no conflict of interest.

Author details

Katarzyna Cesarz-Andraczke*, Aneta Kania, Katarzyna Młynarek and Rafał Babilas
Department of Engineering Materials and Biomaterials, Faculty of Mechanical Engineering, Silesian University of Technology, Gliwice, Poland

*Address all correspondence to: katarzyna.cesarz-andraczke@polsl.pl

References

[1] Braszczyńska-Malik KN. Magnesium alloys and composites based on their matrix. Czestochowa: Czestochowa University of Technology; 2017. 136 p.

[2] Skrzypek SJ, Przybyłowicz K. Engineering of metals and their alloys. Krakow: AGH Publishing House; 2012. 656 p.

[3] Dobrzański LA, Tański T, Dobrzańska-Danikiewicz AD, Król M, Malara S, Domagała-Dubiel J. Structure and properties of Mg-Al-Zn alloys. Open Access Library. International OCSCO World Press; 2012.

[4] Dziubińska A. Forming of flat parts with ribs from magnesium alloy (Ph.D. thesis). Lublin: Lublin University of Technology; 2015.

[5] Marciniak J. Biomaterials engineering. Selected issues. Gliwice: Jacek Skalmierski's Computer Laboratory Publishing House; 2009.

[6] Marciniak J. Biomaterials. Gliwice: Silesian University of Technology; 2013.

[7] Osteoporosis - silent epidemic in Poland [Internet]. 2015. Available from: http://www.osteoporoza.pl/attachments/1907_Raport_System_OP_Minist-7_2015.02.10.pdf [Accessed: 2020-09-14]

[8] 85488 endoprosthesis in 2017 [Internet]. 2017. Available from: https://www.mp.pl/pacjent/ortopedia/aktualnosci/188266,85-488-endoprotezoplastyk-stawowych-w-2017-r?fbclid=IwAR1PRpkyh4u-bYeE4EX4yJfIcsW83Rrw WaOMFk8p8A-leOgltgL5I_3AhW8 [Accessed: 2020-09-14]

[9] Pamuła E. Biomaterials for tissue engineering. Research on the structure and biological properties of aliphatic polyesters. Krakow: Polish Society for Biomaterials; 2008.

[10] Błażewicz S, Stoch L. Biomaterials. Vol 4. Warsaw: Exit Academic Publishing House; 2003.

[11] Greenberg AM, Prein J. Craniomaxillofacial Reconstructive and Corrective Bone Surgery. Principles of Internal Fixation Using the AO/ASIF Technique. Switzerland: Springer-Verlag; 2002.

[12] Babilas R, Bajorek A, Włodarczyk P, Łoński W, Babilas D. Effect of Au addition on the corrosion activity of Ca-Mg-Zn bulk metallic glasses in Ringer's solution. Materials Chemistry and Physics. 2019;226:51-58. DOI: doi.org/10.1016/j.matchemphys.2018.12.088

[13] Karmańska A, Stańczak A, Karwowski B. Magnesium - the actual state of knowledge. Bromatology and Toxicological Chemistry. 2015;4:677-689.

[14] Kabata-Pendias A, Pendias H. Biogeochemistry of trace elements. Warsaw: PWN Publishing House; 1999.

[15] Zhang BP, Wang Y, Geng L. Biomaterials - Physics and Chemistry, Croatia: InTech; 2011.

[16] Vitamin and mineral requirements in human nutrition [Internet]. 1998. Available from: https://apps.who.int/iris/bitstream/handle/10665/42716/9241546123.pdf [Accessed: 2020-09-14]

[17] Reginster JY, Strause L, Deroisy R, Lecart MP, Saltman P, Franchimont P. Preliminary report of decreased serum magnesium in postmenopausal osteoporosis. Magnesium. 1989;8:106-109.

[18] Nechifor M. Magnesium in major depression. Magnesium Research. 2009;22:163S–166S

[19] Guerrera MP, Volpe SL, Mao JJ. Therapeutic uses of magnesium. American Family Physician. 2009;80:157-162

[20] Seal CK, Vince K, Hodgson MA. Biodegradable surgical implants based on magnesium alloys - a review of current research. Materials Science Engineering. 2009;4:012011. DOI: 10.1088/1757-899X/4/1/012011

[21] Chakraborty Banerjee P, Al-Saadi S, Choudhary L, Eslami Harandi S, Singh R. Magnesium implants: prospects and challenges. Materials. 2019;12/1:136. DOI: 10.3390/ma12010136

[22] Agarwal S, Curtin J, Duffy B, Jaiswal S. Biodegradable magnesium alloys for orthopaedic applications: A review on corrosion, biocompatibility and surface modifications. Materials Science and Engineering C. 2016;68:948-963. DOI: 10.1016/j.msec.2016.06.020

[23] Staiger MP, Pietak AM, Huadmai J, Dias G. Magnesium and its alloys as orthopedic biomaterials: a review. Biomaterials. 2006;27/9:1728-1734. DOI: 10.1016/j.biomaterials.2005.10.003

[24] Witte F, Hort N, Vogt C, Cohen S, Kainer KU, Willumeit R, Feyerabend F. Degradable biomaterials based on magnesium corrosion. Current Opinion Solid State Materials Science. 2008;12/5-6:63-72. DOI: 10.1016/j.cossms.2009.04.001

[25] Zheng Y. Magnesium alloys as degradable biomaterials. Boca Raton: CRC Press; 2015.

[26] Ali M, Hussein MA, Al-Aqeeli N. Magnesium-based composites and alloys for medical applications: A review of mechanical and corrosion properties. Journal of Alloys and Compounds. 2019;792:1162-1190. DOI: 10.1016/j.jallcom.2019.04.080

[27] Hong K, Park H, Kim Y, Knapek M, Minarik P, Mathis K, Yamamoto A, Choe H. Mechanical and biocorrosive properties of magnesium-aluminum alloy scaffold for biomedical applications. Journal of Mechanical Behavior and Biomedical Materials. 2019;98:213-224. DOI: 10.1016/j.jmbbm.2019.06.022

[28] Bommala VK, Krishna MG, Rao CT. Magnesium matrix composites for biomedical applications: A review, Journal of Magnesium Alloys. 2019;7(1):72-79. DOI: 10.1016/j.jma.2018.11.001

[29] Echeverry-Rendon M, Allain JP, Robledo SM, Echeverria F, Harmsen MC. Coatings for biodegradable magnesium-based supports for therapy of vascular disease: A general view. Materials Science and Engineering C. 2019;102:150-163. DOI: 10.1016/j.msec.2019.04.032

[30] Chen J, Tan L, Yu X, Etim IP, Ibrahim M, Yang K. Mechanical properties of magnesium alloys for medical application: A review. Journal of the Mechanical Behavior of Biomedical Materials. 2018;87:68-79. DOI: 10.1016/j.jmbbm.2018.07.022

[31] Seitz JM, Lucas A, Kirschner M. Magnesium-based compression screws: a novelty in the clinical use of implants. The Journal of The Minerals, Metals & Materials Society. 2016;68:1177-1182. DOI: 10.1007/s11837-015-1773-1

[32] Ilnicka M, Wawrzyńska M, Biały D. Biodegradable coronary stents – overview. Acta BioOptica et Informatica Medica. 2009;15;369-372. DOI: polona.pl/item/37787964

[33] Lobodzinski SS. Bioabsorbable coronary stents. Folia Cardiologica Excerpta. 2009;4:247-250

[34] Barlis P, Tanigawa J, Di Mario C. Coronary bioabsorbable magnesium

stent: 15-month intravascular ultrasound and optical coherence tomography findings. European Heart Journal. 2007;28:2319-2325. DOI: 10.1093/eurheartj/ehm119

[35] Pekguleryuz M, Kainer K, Kaya A. Fundamentals of magnesium alloy metallurgy. Cambridge: Woodhead Publishing Limited; 2013.

[36] Rokhlin LL, Nikitina NI. Recovery after ageing of Mg-Y and Mg-Gd alloys. Journal of Alloys and Compounds. 1998;279/2:166-170. DOI: 10.1016/S0925-8388(98)00655-0

[37] Kania A, Nowosielski R, Gawlas-Mucha A, Babilas R. Mechanical and corrosion properties of Mg-based alloys with Gd addition. Materials. 2019;12/11:1775. DOI: 10.3390/ma12111775

[38] Hanzi AC, Gerber I, Schinhammer M, Loffler JF, Uggowitzer PJ. On the in vitro and in vivo degradation performance and biological response of new biodegradable Mg-Y-Zn alloys. Acta Biomaterialia. 2010;6/5:1824-1833. DOI: 10.1016/j.actbio.2009.10.008

[39] Heublein B, Rohde R, Kaese V, Niemeyer M, Hartung W, Haverich A. Biocorrosion of magnesium alloys: a new principle in cardiovascular implant technology? Heart. 2003;89/6:651-656. DOI: 10.1136/heart.89.6.651

[40] Cesarz-Andraczke K, Nowosielski R, Babilas R. Corrosion properties of Mg-Zn-Ca-(Cu,Au) metallic glasses in artificial physiological fluid. Archives of Civil and Mechanical Engineering. 2019;19/3:716-723. DOI: 10.1016/j.acme.2019.02.008

[41] Wang L, Qin G, Sun S, Ren Y, Li S. Effect of solid solution treatment on in vitro degradation rate of as-extruded Mg-Zn-Ag alloys. Transactions of Nonferrous Metals Society of China. 2017;27/12:2607-2612. DOI: 10.1016/S1003-6326(17)60288-7

[42] Feng Y, Zhu S, Wang L, Chang L, Guan S. Fabrication and characterization of biodegradable Mg-Zn-Y-Nd-Ag alloy: Microstructure, mechanical properties, corrosion behavior and antibacterial activities. Bioactive Materials. 2018;3/3:225-235. DOI: 10.1016/j.bioactmat.2018.02.002

[43] Nowosielski R, Cesarz K, Nawrat G, Maciej A, Babilas R. Corrosion tests of amorphous and crystalline Mg36.6Cu36.2Ca27.2 alloy in physiological fluid. Corrosion Protection. 2013;4:144-151

[44] Laws K, Shamlaye K, Wong K, Gun B, Ferry M. Prediction of glass-forming compositions in metallic systems: Copper-based bulk metallic glasses in the Cu-Mg-Ca system. Metallurgical and Materials Transactions. 2010;41A:1699-1705. DOI: 10.1007/s11661-010-0274-7

[45] Gu X, Zheng Y, Zhong S, Xi T, Wang J. Wang W. Corrosion of, and cellular responses to Mg-Zn-Ca bulk metallic glasses. Biomaterials. 2010;31:1093-1104. DOI: 10.1016/j.biomaterials.2009.11.015

[46] Li Q, Weng H, Suo Z, Ren Y, Yuan X, Qiu K. Microstructure and mechanical properties of bulk Mg-Zn-Ca amorphous alloys and amorphous matrix composites. Materials Science and Engineering A. 2008;487:301-308. DOI: 10.1016/j.msea.2007.10.027

[47] Park JB, Bronzino JD. Biomaterials: principles and applications. Florida: CRC Press; 2003.

[48] Ding Y, Wen C, Hodgson P, Li Y. Effects of alloying elements on the corrosion behavior and biocompatibility of biodegradable magnesium alloys: A review. Journal of Materials

Chemistry B. 2014;2:1912-1933. DOI: 10.1039/c3tb21746a

[49] Meagher P, O'Cearbhaill ED, Byrne JH, Browne DJ. Bulk metallic glasses for implantable medical devices and surgical tools. Advanced Materials. 2016;28:5755-5762. DOI: 10.1002/adma.201505347

[50] Mushahary D, Sravanthi R, Li Y, Kumar MJ, Harishankar N, Hodgson PD, Wen C, Pande G. Zirconium, calcium, and strontium contents in magnesium based biodegradable alloys modulate the efficiency of implant-induced osseointegration. International Journal of Nanomedicine. 2013;8:2887-2902. DOI: 10.2147/IJN.S47378

[51] Pogorielov M, Husak E, Solodivnik A, Zhdanov S. Magnesium-based biodegradable alloys: Degradation, application, and alloying elements. Interventional Medicine and Applied Science. 2017;9:27-38. DOI: 10.1556/1646.9.2017.1.04

[52] Mostaed E, Vedani M, Hashempour M, Bestetti M. Influence of ECAP process on mechanical and corrosion properties of pure Mg and ZK60 magnesium alloy for biodegradable stent applications. Biomatter. 2014;4:e28283. DOI: 10.4161/biom.28283

[53] Liu C, Ren Z, Xu Y, Pang S, Zhao X, Zhao Y. Biodegradable magnesium alloys developed as bone repair materials: A review. Scanning. 2018;2018. DOI: 10.1155/2018/9216314

[54] Degradation testing of magnesium and its alloys aiming at biodegradable implant applications [Internet]. 2016. Available from: https://lirias.kuleuven.be/bitstream/123456789/552754/1/PhD+thesis+manuscript+-+I%C3%B1igo+Marco.pdf [Accessed: 2020-09-14]

[55] Babilas, R.ł., Cesarz-Andraczke, K., Babilas, D., Simka, W. Structure and Corrosion Resistance of $Ca_{50}Mg_{20}Cu_{30}$ Bulk Metallic Glasses. Journal of Materials Engineering and Performance, 2015; 24/1: 167-174. DOI: 10.1007/s11665-014-1308-x

[56] Schweitzer PA. Fundamentals of metallic corrosion. Atmospheric and media corrosion of metals. New York: CRC Press; 2007.

[57] Rad B, Idris MH, Kadir MRA, Farahany S, Fereidouni A, Yahya MY. Characterization and corrosion behavior of biodegradable Mg-Ca and Mg-Ca-Zn implant alloys. Applied Mechanics and Materials. 2012;121-126:568-572. DOI: 10.4028/www.scientific.net/AMM.121-126.568

[58] Haynes WM. Handbook of Chemistry and Physics. 97th ed. CRC Press; 2016. 2670 p.

[59] Cesarz-Andraczke K. Resorbable Mg-based bulk metallic glasses [Ph.D. thesis]. Gliwice: Silesian University of Technology; 2016.

[60] Wu W, Wang Z, Zang S, Yu X, Yang H, Chang S. Research progress on surface treatments of biodegradable Mg Alloys: A Review. ACS Omega. 2020;5:941-947. DOI: 10.1021/acsomega.9b03423

[61] Li Y, Zhao S, Li S, Ge Y, Wang R, Zheng L, Xu J, Sun M, Jiang Q, Zhang Y, Wei H. Surface engineering of biodegradable magnesium alloys for enhanced orthopedic implants. Nano Micro Small. 2019;15:1-10. DOI: 10.1002/smll.201904486

[62] Hornberger H, Virtanen S, Boccaccini AR. Biomedical coatings on magnesium alloys - A review. Acta Biomaterialia. 2012;8:2442-2455. DOI: 10.1016/j.actbio.2012.04.012

[63] Uddin MS, Hall C, Murphy P. Surface treatments for controlling corrosion rate of biodegradable

Mg and Mg-based alloy implants. Science and Technology of Advanced Materials. 2015;16:53501. DOI: 10.1088/1468-6996/16/5/053501

[64] Zeng RC, Yin ZZ, Chen XB, Xu DK. Corrosion types of magnesium alloys. In: Magnesium Alloys - Selected Issue [Internet]. 2018. Available from: https://www.intechopen.com/books/magnesium-alloys-selected-issue/corrosion-types-of-magnesium-alloys [Accessed: 2020-09-14]

[65] Zhang P, Zuo Y. Relationship between porosity, pore parameters and properties of microarc oxidation film on AZ91D magnesium alloy. Results in Physics. 2019;12:2044-2054. DOI: 10.1016/j.rinp.2019.01.095

[66] Zheng YF, Gu XN, Witte F. Biodegradable metals. Materials Science and Engineering: R: Reports. 2014;77:1-34. DOI: 10.1016/j.mser.2014.01.001

[67] Echeverry-Rendon M, Allain JP, Robledo SM, Echeverria F, Harmsen MC. Coatings for biodegradable magnesium-based supports for therapy of vascular disease: A general view, Materials Science and Engineering C. 2019;102:150-163. DOI: 10.1016/j.msec.2019.04.032

[68] Gao Y, Jie M, Liu Y. Mechanical properties of Al_2O_3 ceramic coatings prepared by plasma spraying on magnesium alloy. Surface and Coatings Technology. 2017;315:214-219. DOI: 10.1016/j.surfcoat.2017.02.026

[69] Jiang S, Cai S, Lin Y, Bao X, Xu G. Effect of alkali/acid pretreatment on the topography and corrosion resistance of as-deposited CaP coating on magnesium alloys. Journal of Alloys and Compounds. 2019;793:202-211. DOI: 10.1016/j.jallcom.2019.04.198

[70] Feng Y, Zhu S, Wang L, Chang L, Yan B, Song X, Guan S. Characterization and corrosion property of nano-rod-like HA on fluoride coating supported on Mg-Zn-Ca alloy. Bioactive Materials. 2017;2:63-70. DOI: 10.1016/j.bioactmat.2017.05.001

[71] Shi P, Niu B, Shanshan E, Chen Y, Li Q. Preparation and characterization of PLA coating and PLA/MAO composite coatings on AZ31 magnesium alloy for improvement of corrosion resistance. Surface and Coatings Technology. 2015;262:26-32. DOI: 10.1016/j.surfcoat.2014.11.069

[72] Manna S, Donnell AM, Kaval N, Marwan F. Improved design and characterization of PLGA/PLA-coated Chitosan based micro-implants for controlled release of hydrophilic drugs. International Journal of Pharmaceutics. 2018;547/1-2:122-132. DOI: 10.1016/j.ijpharm.2018.05.066

[73] Reza H, Rad B, Fauzi Ismail A, Aziz M, Hadisi Z, Chen X. Antibacterial activity and corrosion resistance of Ta_2O_5 thin film and electrospun PCL/MgO-Ag nanofiber coatings on biodegradable Mg alloy. Ceramics International. 2019;45/9:11883-11892. DOI: 10.1016/j.ceramint.2019.03.071

[74] Ling L, Cui L, Zeng R, Li S, Chen X, Zheng Y, Kannan MB. Advances in functionalized polymer coatings on biodegradable magnesium alloys - A review. Acta Biomaterialia. 2018;79:23-36. DOI: 10.1016/j.actbio.2018.08.030

[75] Mashtalyar DV, Sinebryukhov SL, Imshinetskiy IM, Gnedenkov AS, Nadaraia KV, Ustinov AY, Gnedenkov SV. Hard wearproof PEO-coatings formed on Mg alloy using TiN nanoparticles. Applied Surface Science. 2020;503:1-12. DOI: 10.1016/j.apsusc.2019.144062

[76] Wang S, Si N, Xia Y, Liu L. Influence of nano-SiC on microstructure and property of MAO coating formed on AZ91D magnesium alloy. Transactions of Nonferrous Metals Society of China.

2015;25:1926-1934. DOI: 10.1016/S1003-6326(15)63800-6

[77] Singh B, Singh G, Sidhu BS, Bhatia N. In-vitro assessment of HA-Nb coating on Mg alloy ZK60 for biomedical applications. Materials Chemistry and Physics. 2019;231/1:138-149. DOI: 10.1016/j.matchemphys.2019.04.037

[78] Cesarz-Andraczke K, Nowosielski R, Basiaga M, Babilas R. Study of the morphology and properties of biocompatible Ca-P coatings on Mg alloy. Materials. 2020;13/2:1-13. DOI: 10.3390/ma13010002

[79] Sobczak-Kupiec A, Wzorek Z. Physicochemical properties of calcium orthophosphates important for medicine. Chemistry. 2007;10:309-322.

Chapter 5

Mg-Based Composites for Biomedical Applications

Moara Marques de Castro, Débora Ribeiro Lopes and Leonardo Viana Dias

Abstract

Magnesium (Mg) is a promising material for producing temporary orthopedic implants, since it is a biodegradable and biocompatible metal which density is very similar to that of the bones. Another benefit is the small strength mismatch when compared to other biocompatible metals, what alleviates stress-shielding effects between bone and the implant. To take advantage of the best materials properties, it is possible to combine magnesium with bioactive ceramics and tailor composites for medical applications with improved biocompatibility, controllable degradation rates and the necessary mechanical properties. To properly insert bioactive reinforcement into the metallic matrix, the fabrication of these composites usually involves at least one high temperature step, as casting or sintering. Yet, recent papers report the development of Mg-based composites at room temperature using severe plastic deformation. This chapter goes through the available data over the development of Mg-composites reinforced with bioactive ceramics, presenting the latest findings on the topic. This overview aims to identify the major influence of the processing route on matrix refinement and reinforcement dispersion, which are critical parameters to determine mechanical and corrosion properties of biodegradable Mg-based composites.

Keywords: magnesium, biodegradable composite, bioactive material, temporary implant, severe plastic deformation

1. Introduction

It is known that magnesium is a very low-density metal with great importance for structural applications where low weight is desirable. It is approximately 35% lighter than aluminum, 60% lighter than titanium and 78% lighter than steel. Magnesium and its alloys are also potential materials for structural orthopedic implants, since not only Mg's density (1.738 g/cm^3) but also its tensile strength (80–280 MPa) are similar to that of the bone, with advantages as higher fracture toughness [1]. The similar density and smaller elastic modulus mismatch when compared to other biocompatible metals as stainless steel or titanium alloys alleviates stress-shielding effects between bone and the implant material [2].

For being highly biocompatible and biodegradable, Mg is a potential material for producing temporary implants, dispensing a second surgery to remove the implant after the damage tissue is completely healed. Despite their high biocompatibility and no toxic risk during degradation, the corrosion rate of Mg is still too fast, compromising the structural function of the medical device.

IntechOpen

Many studies have been conducted over the topic to overcome these issues, and among the potential solutions, the fabrication of composites show promising results. Some reasons for choosing this approach are the possibility to take advantage of bioactive ceramics' effects as assisting healing or improving the bonding between the tissue and the device, together with the benefits of having a metallic matrix, as good mechanical resistance, better ductility and good energy of impact absorption. Besides, a proper processing technique leads to a more uniform corrosion though the composite, lowering the degradation rate.

This chapter presents an overview having some of the newest studies over Mg-based composites for biological applications and provide a comparison among the processing techniques, effect of reinforcement content in corrosion and mechanical properties.

2. Biocompatible metallic materials

Medical parts as implants, stents, scaffolds and fixation devices are used when a bone is broken, for supportive structural purposes, bearing the body's load until the tissue can properly regenerate. These parts must present the following properties:

- Be biocompatible and have maximum cell viability to avoid any harm as inflammation, infection, or other adverse reactions;
- Not be toxic to the body, and not release toxic elements during degradation,
- Have a morphology that improves healing process and osseointegration [3];
- Have high enough corrosion resistance to assure mechanical integrity until the healing of the tissue;
- Remain stable and integer despite the constant impact of body movement;
- Have suitable mechanical properties as high toughness, tenacity, fatigue strength, stiffness comparable elastic modulus to bone to avoid stress shielding effects [4], and sufficient hardness to increase wear resistance and prevent wear debris.

Biomaterials currently used for this application can be polymers, metals, ceramics or composites. Advantage of metals for load-bearing application is the combination of better mechanical strength compared to polymers and better tenacity when compared to ceramics. The most common metals used are titanium alloys, stainless steel and chromium-cobalt alloy, that are bioinert. These metals have high strength to support body load and can keep a good integrity, staying longer in the body due to their high-corrosion resistance [5]. Iron, magnesium, and zinc are also biocompatible metals, with the advantage of being biodegradable in physiological environment. Other metallic biomaterials include shape memory alloys, tantalum and some precious metals [6].

Table 1 summarizes some mechanical properties, density, pros and cons of the usual biocompatible metals [6, 7]. For comparison, the properties of the cortical bones are: a density around 1.8 g/cm^3, a modulus of elasticity of the order of 17.0–20.0 GPa in longitudinal direction and 6.0–13.0 GPa in the transversal direction, the tensile strength in the range of 78.8–151.0 MPa in longitudinal direction and 51.0–56.0 MPa in transversal direction, and a fracture toughness in the range of 2–12 MPa.m$^{1/2}$ [8].

Material	Stainless steel	Cobalt-base alloys	Ti and alloys	Mg and alloys
Density [$g.cm^{-3}$]	7.4–8	8–8.5	4.5	1.74–1.81
Young's modulus [GPa]	200	230	106	45
Ultimate tensile strength [MPa]	540–1000	900–1540	900	80–280
Fracture toughness [$MPa.m^{1/2}$]	~100	~100	~80	~15
Advantages	Cost, availability, processing	Wear resistance, corrosion resistance, fatigue strength	Biocompatibility, corrosion, low modulus, fatigue strength	Biocompatibility, biodegradable, low modulus, low density
Disadvantages	Long-term behavior, high modulus	High modulus, biocompatibility	Lower wear resistance, low shear strength	High hydrogen evolution during degradation
Primary uses	Temporary devices (fracture plates, screws, hip, nails)	Dentistry casting, prostheses stems	Femoral heads, long-term permanent devices (nails, pacemakers)	Bone screws, plates, pins, stents, nails

Table 1.
Characteristics of usual metallic implant materials. Values taken from Ref. [6].

It is possible to notice a considerable mismatch of mechanical properties of these bioinert metals and the cortical bones. When the stress transfer between the implant and the tissue is not homogeneous, stress shielding of the bone may occurs. This takes place by differences between Young's moduli of the implant and the bone, what may interfere the regeneration process, cause pain, bone atrophy with resorption of the bone surrounding the implant, loosening of the implant and possibly inducing a new fracturing [9]. Almost 10% from overall operations would undergo for revision surgery for replacing a previously implant, and 79% of all revisions were due to implant loosening [10]. The most common cause of implant loosening is the loss of bone mass due to stress shielding [10].

In the cases which the tissue can regenerate, the implant loses its function after the healing process, and additional surgery may be required to remove it. Besides, an implant which stay in the body for a long period of time can induce some complications, i.e. release toxic elements, permanent physical irritation, a chronic inflammatory response, or infection caused by debris and corrosion products [11, 12].

Biodegradable implants seem to be a clever choice for application where the body can eventually regenerate, since this would dispense the need of a secondary surgery to remove it, diminishing risks of inflammation, pain, surgery risks and costs [2, 12]. So, it is of great interest to develop devices which can be gradually degraded and naturally absorbed by the body while assist tissue healing. Yet, it has been challenging to obtain the combination of a controllable degradation rate, good biocompatibility, and suitable mechanical properties.

3. Magnesium as a biodegradable material

Among the available biodegradable metals, magnesium is very advantageous since it offers necessary stiffness, toughness and tenacity, high damping capacity, very low density and high biocompatibility. It is the fourth most abundant mineral in the body, being present mostly as surface substituents of the bone's hydroxyapatite, and in skeletal muscle and soft tissue [13]. Also, Mg is involved in more than 300 essential metabolic reactions, being a cofactor for several enzymes that stabilize RNA and DNA structures. It is necessary to assist in cardiac, muscular, nervous, bone and renal function [14]. Its ion (Mg^{2+}) that is released during degradation is used in the regular metabolism having no critical toxic limits or side effects reported [3, 13]. It has great ability to be absorb and excreted by the body without causing any harm (for example, the corrosion product MgO is totally removed in the urine [11]). During the first half of the 20th century, magnesium was used in surgeries to connect blood vessels and, in some cases, portions of the intestine in animals and in humans. The most advanced clinical applications are biodegradable cardiovascular magnesium stents; however, because it is bioabsorbable, research has been carried out aiming orthopedic applications as non-permanent implants.

The biggest issue that hinders the use of Mg as a biomedical material is its too fast corrosion in body fluid. This characteristic can be related to three main factors: Mg alloys are very reactive, the corrosion products formed in aqueous solutions containing chloride is not protective, and the micro-galvanic interaction between the constituent phases causes the corrosion of the Mg alloys to be significantly greater than that of the Mg of high purity [15].

Besides an early loss their structural function, a fast corrosion releases a big quantity of hydrogen, what may induce some local pH changes, affecting some physiological reactions and leading to an alkaline poisoning effect. Moreover, gas bubbles can accumulate in the implants surrounding surface delay healing and causing necroses due tissue detachments in some cases [3, 16].

3.1 Mg corrosion

Two types of corrosion occur preferentially in Mg and its alloys, galvanic corrosion and localized corrosion (by pitting or filiform). Galvanic corrosion occurs because magnesium generally behaves anodically in contact with other metals, and it is often used as a sacrificial anode. The standard reduction potential of Mg to form Mg^{2+} is $E° = -2.37$ VSHE and one of its main alloy elements is aluminum (Al), which has a reduction potential $E° = -1.66$ VSHE in the reaction reduction for Al^{3+}. Galvanic pairs are generally formed when the concentration of the alloying element exceeds the maximum solubility limit. The extent of the galvanic effect depends on several factors, such as the crystalline orientation of the magnesium matrix, the type of secondary phases, particles of impurities, the size of the grain and the medium.

The concentration and distribution of the secondary phases are also important. A fine and continuous distribution of the secondary phases normally increases the corrosion resistance of the metal in various media. Alloy elements and other intermetallic phases formed, with electrochemical potential close to that of magnesium, can increase corrosion resistance, reducing galvanic internal corrosion [16].

Magnesium, when immersed in an aqueous solution, presents a double oxide film, an internal layer of MgO, which is adhered to the metal surface and an external and porous layer of Mg $(OH)_2$ [17]. MgO oxide is a n-type semiconductor, with ΔG formation of −136 kcal/mol at 25°C, has a PB (Pilling Bedworth) ratio of less than 1, and is therefore only partially protective, since the volume of the oxide formed is less than the volume of the reagent metal. The Mg $(OH)_2$ film has a lamellar structure that facilitates basal cleavage, and has a PB ratio of 1.77, which generates internal compressive stress, leading to cracks in the $Mg(OH)_2$ film.

In the case of magnesium, unlike aluminum or steel, the oxide layer is crystalline. There is no epitaxy between the oxide layer and the matrix with the compact hexagonal mesh (HCP), leading to a high compressive stress of the layer [18]. One of the ways to reduce discontinuity and have less disorder between the oxide layer and the metal surface is to introduce a large fraction of grain outlines per unit area [19].

Cathodic reactions in neutral or alkaline aqueous media are oxygen reduction to hydrogen evolution, expressed in Eq. 1:

$$H_2O_{(l)} + 2\,e^- \rightarrow H_{2(g)} + 2\,OH^-_{(aq)} \tag{1}$$

In order to have a complete understanding of the Mg corrosion mechanism, it is necessary to consider a phenomenon called Negative Difference Effect (NDE). This phenomenon is experimentally characterized by an unexpected increase in the hydrogen evolution reaction when the anodic overvoltage is increased. NDE is the main cathodic reaction in the corrosion of Mg and occurs spontaneously in Mg under open circuit conditions. When Mg and its alloys are anodically polarized, hydrogen evolution also occurs, which is normally restricted to the cathodic branch. Anodic polarization causes higher rates of dissolution, and higher rates of NDE are also observed [20]. To explain the corrosion of magnesium and more particularly the NDE phenomenon, several mechanisms have been proposed including the formation of magnesium hydrides, metastable monovalent ions, hydroxides and magnesium oxides [17].

The corrosion rate of magnesium-based biomaterials, when implanted in the human body, depends on the composition of the material and the environmental conditions in which the implant is found, such as temperature, pH and concentration of other ions. A study [21] indicates that, the effects of Mg corrosion products on cellular activity is concentration-dependent, on bone marrow derived stem cells and on osteoclastogenesis *in vitro*. This will vary depending on the state of differentiation of cells and length of exposure. Also, the presence of the corrosion products significantly altered the cells' metabolic and proliferative activities, which further affected cell fusion/differentiation. Maradze and coauthors [21] summarizes the cellular response in the presence of a corroding Mg biomaterial *in vivo* in **Figure 1**.

The corrosion rate of Mg can be reduced by increasing the intensity of the basal planes parallel to the surface [22]. Other strategies to diminishing the degradation rate are grain refinement, alloying, surface modification, conversion coating, and incorporation of other materials to create composites. It is worth noting that, in the case of composites, some of these strategies can be combined. For example, a processing technique that enables a refined microstructure can be selected. Grain refinement has a recognized importance for improving mechanical resistance and also for diminishing the corrosion rate with time due to the development of a more homogeneous protective layer on composites surface [23–25]. Also, the matrix can either be pure Mg or an alloy. Although many Mg-alloys are been developed for improved corrosion resistance and superior mechanical strength [4, 11, 26], care must be taken to choose a composition free of any toxic elements. The insertion of a biocompatible second phase in a magnesium matrix could bring a better mechanical resistance and less pitch corrosion. Furthermore, some bioactive materials that could promote a better bonding between implant and the tissue can be incorporated and accelerate body's healing response in some cases. Also, the addition of a hard phase, combined with mechanical processing, enables a better grain refinement and strengthening. Thus, magnesium matrix composites are potential candidates for structural orthopedic implants.

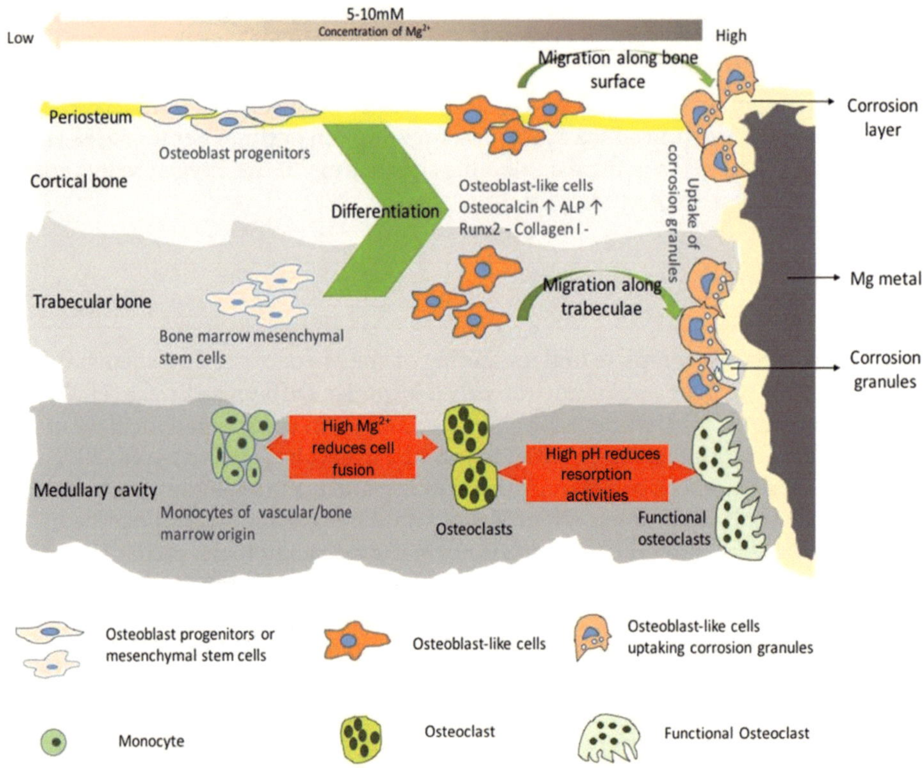

Figure 1.
Representation of cellular response in the presence of a corroding Mg biomaterial in vivo [21]. Scientific reports – Open access.

4. Magnesium-based composites for biomedical applications

There are three determinant factors for the biocompatibility, mechanical strength and corrosion behavior of the Mg-composite: grain size, second phase distribution, and materials composition. The first two can be controlled by fabrication and processing technique. The deformation route directly influences the texture of the material, that has a great influence on magnesium corrosion. Also, the reinforcement selection (nature, fraction and morphology) is of great importance for dictating the composite's properties, and it will determine if the device is toxic, inert, or bioactive.

4.1 Selecting the magnesium-matrix

Elements commonly used for Mg alloying are Al, Ca, Cu, Fe, Li, Mn, Ni, Sr., Y, Zn, Zr, and rare earth elements.

Aluminum is the most usual alloying element for Mg. It can enhance the alloy strength by both solid solution and precipitation of intermetallics. Another advantage is the low density of this metal (2.7 g/cm^3). Although alloys as AZ91 and ZK60 have been used for biomedical applications, this element should be avoided due to the risk factor of Alzheimer's disease, muscle damage, decrease of activities of osteoclasts, altered functions of the blood–brain-barrier, and increase estrogen-related gene expression in human breast cancer cells when cultured in a laboratory setting [11, 27].

Manganese can improve Mg corrosion resistance by reducing the harmful effects of impurities [27]. Mn is also an essential biological trace mineral that acts in many cellular systems, especially as cofactors for many metalloenzymes as oxidases and dehydrogenases, DNA and RNA polymerases, kinases, decarboxylases and sugar transferases, although excessive amounts could induce neurological disorder [27].

Zinc is another common alloying element in Mg. It is nontoxic, biocompatible and biodegradable, and plays a significant role in human body. Zn is present in metabolic activities, as co-factor for some enzymes and it is essential for immune system. It is readily absorbable by biological functions within the cell [27]. The addition of until 4% wt. Zn increase ultimate tensile strength and elongation of as-cast Mg-Zn alloys.

Calcium is the most abundant mineral in the human body, important or bone function, vascular and heart physiology. Adding in small amounts, it can contribute for mechanical strengthening by solid solution, precipitation and grain boundary pinning. Mg_2Ca is brittle and the addition of >1%Ca deteriorate ductility and mechanical properties. Despite Ca positively influence the cell viability and proliferation rate, the effect of Ca in corrosion resistance is deleterious because it accelerates degradation due to galvanic corrosion.

Kirkland et al. [28] perform a systematic study comparing the degradation of a series of Mg-Ca-Zn alloys. **Figure 2** present average values of the maximum compression strength of their Mg-xCa and Mg-xZn alloys after 1 and 3 weeks of immersion in SBF, in relation to the alloying content. The blue lines indicate the compressive strength of pure Mg, after 3 weeks in SBF (lower line) or without immersion (upper line).

An issue related to the most common alloys is concerned about the biocompatibility and toxicity of alloying elements [29]. The alloying elements might me selected to not only improve corrosion and mechanical properties of the alloy, but also to improve the body response. But mostly, special care must be taken to avoid elements that may release corrosion products that can be toxic to the body. A summary of the influence of metal ions on the variety of processes involved in bone regeneration is depicted in **Figure 3** [30].

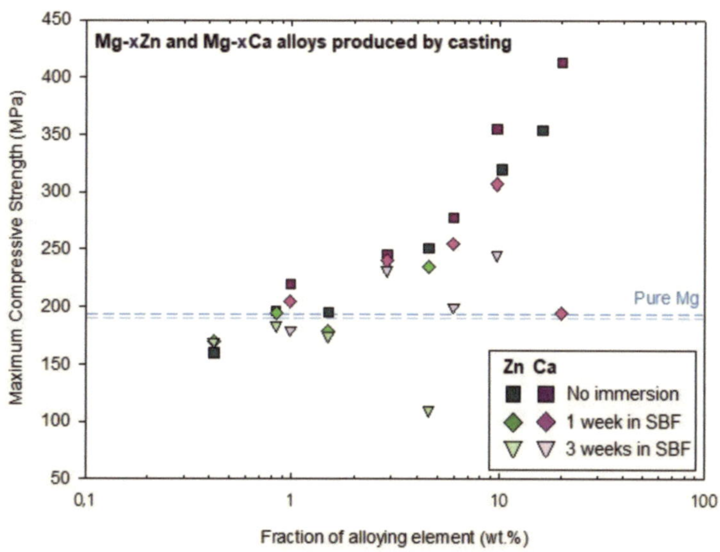

Figure 2.
Effect of alloying content in the compression strength of Mg-xCa and Mg-xZn alloys after 1 and 3 weeks immersed in SBF. Based on Ref. [28].

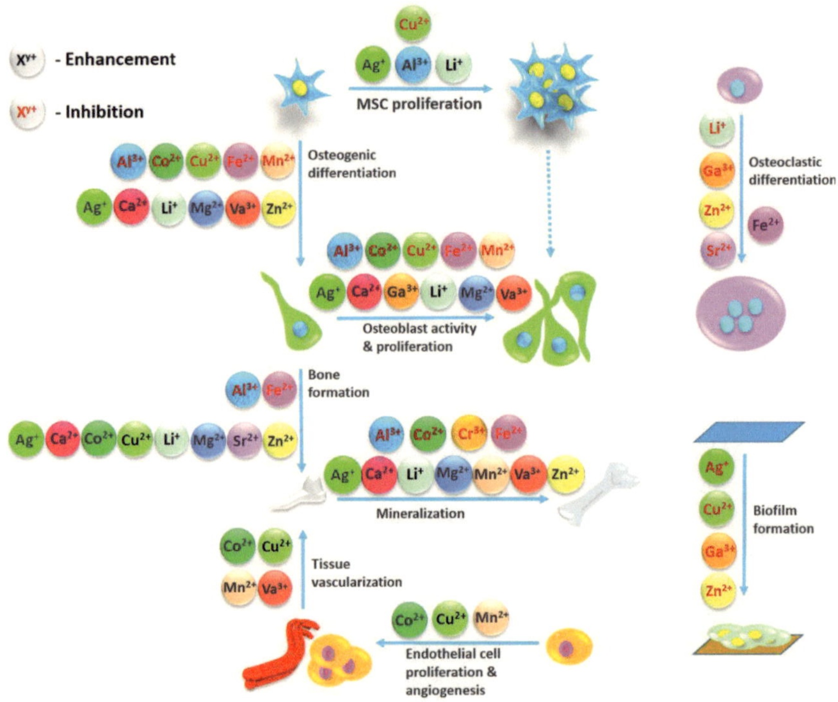

Figure 3.
Summary of the influence of metal ions on processes involved in bone regeneration [30]. MDPI-open access.

4.2 Selecting the reinforcement

The reinforcement selection (nature, fraction and morphology) is of great importance for dictating the composite's properties. Primarily, it should be highly compatible and safely absorbable by the body without any harm to the organism. It should also achieve a good interfacial bonding with the matrix, since a discontinuous matrix can lead to stress concentration, facilitating crack evolution and increasing localized corrosion.

The most common materials used to reinforce Mg-based composites aiming biomedical applications are silica-based or phosphate-based ceramics [4]. Bioactive glasses (BG) is an important silica-based material that can make a strong bonding to the bone. It is intrinsically brittle. They display significant bioactivity, improves hemocompatibility, and the ionic dissolution products of bioactive glasses stimulate osteoblast proliferation [31]. They are used in dental implants and for artificial bone. Some papers have reported that addition of BG to Mg matrix can improve the biocompatibility of pure Mg [32]. Yet, calcium phosphates such as hydroxyapatite (HA) and tricalcium phosphate (TCP) are the most popular choice to fabricate Mg-matrix composites aiming biomedical application. B-TCP has a great importance for being bioresorbable, bioactive and osteoconductive material. It is used for bone tissue regeneration. HA can form strong chemical bonds with the osseous tissue and it is one of the main components affecting the mechanical strength to bone, providing it stiffness [1]. It is commonly used as orthopaedical implants, dental implants and coating metallic implants, tissue engineering scaffolds. Yet, bone grafts made exclusively of HA present a low fracture toughness (0.7 MPa.m$^{1/2}$, [16]).

The opportunity to match the superior compressive strength and biological performance of bioactive ceramics, with the toughness and resilience of Mg and production of biodegradable and bioactive composites is of great research interest.

There are many researches over the fabrication of a magnesium-based composite with biocompatible ceramic reinforcements dispersed into the metallic matrix. Many review papers about Mg-based composites for biomedical applications have also been published, covering since fabrication methods, mechanical properties improvement, corrosion behavior and also biocompatibility *in vitro* and *in vivo* [4, 7, 33–35].

4.3 Selecting the fabrication technique

The fabrication methods used to produce Mg-based composites for biomedical purposes are listed below. The processes techniques can be divided in either liquid state (casting) or solid-state processing route (powder metallurgy). Routes that involves casting promotes the insertion of the hard phase in a liquid bath of the matrix-metal, while solid-state routes achieve dense consolidated composites without melting the materials. In this case, the particles consolidation should be induced by high-temperature diffusion and/or plastic deformation. To ensure a full densification of the composite, great majority of the solid-state techniques takes place at high temperatures, but below materials' melting point. All processing techniques should enable an efficient matrix-reinforcement bonding and a homogeneous dispersion of the hard phase, since agglomeration and internal cracks are deleterious to corrosion and mechanical properties [36, 37].

4.3.1 Casting

Many metallic composites are synthesized by casting techniques. They can be fabricated by pumping the melted metal at high pressure into a die which can contain the reinforcement dispersed or as a preform. Alternatively, the hard phase can be added to the matrix by stirring it into the melted metallic bath.

Khanra et al. used stir casting to fabricated Mg-HA and ZM61-HA composites having different amounts of HA (0, 5, 10 and 15 wt%), then performed hot extrusion at 320°C on the billets [38, 39]. The addition of HA induced grain refinement, an increase in compressive strength but a decrease in tensile strength. Since the alloy have superior strength than the pure metal, (ultimate tensile strength – UTS of the Mg-0 HA is 187.9 while UTS of ZM61-0HA is 301 MPa) the ZM61-HA composites presented higher mechanical properties than Mg-HA ones (UTS of Mg-15HA is 136.7 MPa, against 225.5 MPa from the ZM61-15HA).

Melting and hot extrusion were also applied to synthesize Mg-Zn-Zr matrix composites with 0, 0.5, 1 and 1.5 wt% HA as reinforcement [40, 41]. Common results were the improved of mechanical properties with HA addition, and reduction in corrosion rate for composites having 1 wt% HA.

Kumar et al. [42] synthesized HA-MgO composites using 0, 0.25, 0.5, 1 and 2% MgO by melting infiltration. The mass loss, after 21 days in SBF, of pure HA is 0.43 ± 0.08 mg, which is significantly higher than the HA-1.0 MgO composite (0.21 ± 0.03 mg). As increasing the amounts of MgO, a remarkable increase in the mechanical properties of the composite was achieved.

4.3.2 Power metallurgy processing

Powder metallurgy (PM) techniques involves a methodology composed of powders mixture, compaction and sintering of the materials, usually in an inert atmosphere. The mixture is commonly performed by ball milling especially when a better interaction between matrix and reinforcement is expected. Uniaxial pressing at room temperature is generally applied to create the green compacts before the sintering steps. Other processes as isostatic pressure, hot pressing [43, 44] and double-step sintering [45] have also been used to assure a better densification of Mg-HA composites. Conventional sintering strongly relies on diffusional processes, that are favored by longer times and higher temperatures, and this can lead to a significant grain growth during the process.

Setyadi et al. [46] fabricated Mg-CA composites having 5, 10, 15 wt% of carbonate apatite (CA) by powder metallurgy. The higher milling times and higher content of CA lead to an increase in hardness of until ~20% when compared to pure magnesium (hardness whet from 37.3 HV in pure consolidated Mg to 44.8 HV for 7 h of milling and 15% CA). The optimal composition was obtained for the composite with 10 wt% of CA. Salleh et al. [47] produce Mg6.5Zn matrix composites with 0 and 10 wt% of HA by mechanical milling and PM. The progressive effect of HA followed by the Zn addition, decelerate the degradation rate of Mg. The composite exhibited the highest corrosion resistance in Hank's Balanced Salt solution (HBSS). The compressive strength for the alloy and the composite, after 7 days of immersion, are considerably higher compared to the cortical bone.

Double step sintering PM (450°-550°) was applied by Jaiswal et al. [45] to fabricate Mg-3Zn matrix composite reinforced with 0, 2, 5 and 10 wt% of HA. It was reported that the addition of 5 wt% HA is found effective in reducing the corrosion rate by 42% in SBF. Mg-Zn with 5% HA showed improvement in the compressive yield strength of biodegradable magnesium alloy by 23%.

Ball milling and hot extrusion at 270°C were used by Stüpp et al. [43] to produce ZK60-HA composite with 0, 10, 20 wt% HA. A slight improvement in the corrosion resistance was observed for the composites in DMEM + FBS. Samples showed a slight increase in the compressive yield strength with the addition of HA.

Spark plasma sintering (SPS) apply high-pressure to enable a better densification of the sample using considerably less time and lower temperatures than in conventional sintering, and thus achieves finer grain sizes. Sunil et al. [48] used ball milling and SPS at 450° to produced Mg-HA composites having 0, 8, 10 and 15 wt% of HA. Fracture toughness and Young's modulus decreased with the increase of HA content, but were improved in relation to the as cast Mg. The Mg-10%HA composite present the better corrosion resistance among the samples.

Microwave-assisted processing is another sintering technique which a uniform heat is induced with an accurate energy balance by electromagnetic waves in a high-frequency electric field. It promotes a rapid sintering of the Mg to a sample which density is near the theoretical value, using relatively low energy consumption. This approach was used by Wan et al. [32] to produce Mg-BG composites with 5, 10 and 15% of BG. The materials were ball milled and sintered in a microwave furnace at 500°C. It was reported that the addition of bioactive glass can considerable reduce hydrogen evolution, avoid significant pH change and improve biocompatibility when compared to pure Mg. The best mechanical and biological behavior was presented by the Mg-10%BG composite. The improvements in compressive strength and modulus are 34.0 and 18.7%, respectively, compared to pure Mg. Xiong et al. [49] also used the same procedure to prepare Mg-HA composites having 5, 10 and 15 wt% of HA, and observed that mechanical properties and corrosion resistance of Mg-HA composites were better than the ones of pure Mg. The HA was well dispersed in all samples, but the optimum composition was found to be 10% of HA. This reinforcement addition not only presented the lower corrosion current density ($1x\,10^{-4}$ A/cm^2) but also increased the compressive strength and modulus by about 67.5 and 42.8%, respectively.

In the available review papers, it is possible to notice that all methods described to produce the biodegradable composites involve at least one steep of high temperature processing. Yet, it was recently reported that different bioactive composites were fabricated by high-pressure torsion (HPT) at room temperature, combining Mg and alloys with bioactive reinforcement as hydroxyapatite or bioactive glass [50].

4.3.3 Severe plastic deformation

Severe plastic deformation (SPD) are processes techniques where high hydrostatic pressures are applied while the materials goes through an intensive shear, but without significant change in sample's dimensions. This enables multiple consecutive process steps and then accumulation a great amount of deformation. As a result, the processed materials go through extensive grain refinement, homogenization of second phases' distribution and alteration of the texture, which can contribute to the reduction of corrosion [24, 25, 37, 51]. Yet, it is observed that corrosion analysis of magnesium alloys with ultrafine grains is particularly complex, as it may involve factors such as texture, formation of unstable passive film and of heterogeneous structures resulted from dynamic recrystallization.

Equal channel angular pressing (ECAP) and high-pressure torsion (HPT) are well-known SPD techniques that can not only process bulk samples but also consolidate metal particles and produce metal-matrix composites with a refined microstructure. The mechanism of powder consolidation SPD processes differs from conventional sintering routes because the bonding between particles is caused by deformation of the particles, rather than the atomic diffusion. Thus, shorter times and lower temperatures are needed to consolidate the particles into a fully dense sample.

Silva et al. [25] reported the beneficial effects in SPD Mg in compared to as cast or hot rolled samples. As is showed in **Table 2**, the SPD processed samples present smaller grain sizes, thus better mechanical properties and better corrosion properties. Besides the formation of a protective film composed of corrosion products - that reduced the loss of mass after long intervals immersion – Mg biocompatibility was not affected by this treatment.

It is difficult to standardize and control the texture of the alloys during SPD processing. However, it is known that HPT processing leads to a preferential orientation of the grains, of the magnesium alloy, in (0002) or basal plane. Since the basal plane is more stable than other non-oriented planes or grains, thus the increase in oriented grains (0002) increases corrosion resistance [18].

High-pressure torsion was also used to fabricate Mg-based bioactive composites, but at room temperature [50, 52]. The matrix where composed by commercially pure Mg or the AZ91 alloy, and the selected reinforcement were bioactive glass or hydroxyapatite. SEM images of the materials are depicted in **Figure 4**.

Processing	Grain size (μm)	Yield stress (MPa)	Corrosion potential (V)	Corrosion current (A/cm^2)	Corrosion rate* ($g/cm^2/h$)
As cast	480	34	−1.48	7	54
Hot-rolled	16	110	−1.61	350	—
Hot-rolled + ECAP	3.2	140	−1.51	40	25
HPT	0.56	80	−1.49	8	19

**Corrosion rate calculated from the H_2 evolution after 48 h of exposure in 3.5% NaCl solution.*

Table 2.
Properties of pure Mg processed by different techniques [25].

Figure 4.
Particles used as starting materials to produce Mg-based composites at room temperature using high-pressure torsion [50]. [50] MDPI-open access.

The AZ91–5%BG and Mg-5%BG composites [50] exhibited small impedance arcs and fast corrosion in HBSS due to the presence of cracks developed inside BG particles during processing. Adding the bioactive ceramic improved hardness of all composites, and this gain was more expressive for the ones with CP-Mg matrix (77% higher than the monolithic HPT-processed Mg). It was reported that a good dispersion of HA and an efficient consolidation of the matrix were achieved for the Mg-5HA (**Figure 5**). The tensile strength and the hardness of this composite was improved in relation to the pure metal. As Mg is soft and the particles are relatively big, it can easily deform around the ceramic particles, either HA or BG. On the other hand, AZ91 has a better strength than the pure metal (**Figure 6a**), so it is harder to consolidate. Besides, bioactive glass particles are sharp and fragile, so they can fracture during processing and induce localized corrosion in discontinuities surrounding.

This corroborates with the poor corrosion resistance of the BG reinforced composites, especially the AZ91–5BG, as shown in **Figure 6b**. The size and nature of the reinforcement are indeed determinant for slower corrosion rate due to the tendency to develop cracks during processing. This fact was also observed in Mg-quasicrystals composites [53] which fractured along the coarse quasicrystalline particles and on the matrix-reinforcement interface after tensile test.

In the Mg-5%HA composite [52] the addition of HA enhanced the corrosion resistance after 10 hours of immersion in HBSS, the impedance increases and the mass loss rate decreases in the composite. The composite exhibited enhanced hardness and ultimate tensile strength in comparison to pure Mg process by HPT. The corrosion evolution behavior of the Mg-5%HA reported by Lopes et al. [52] seems to be slower

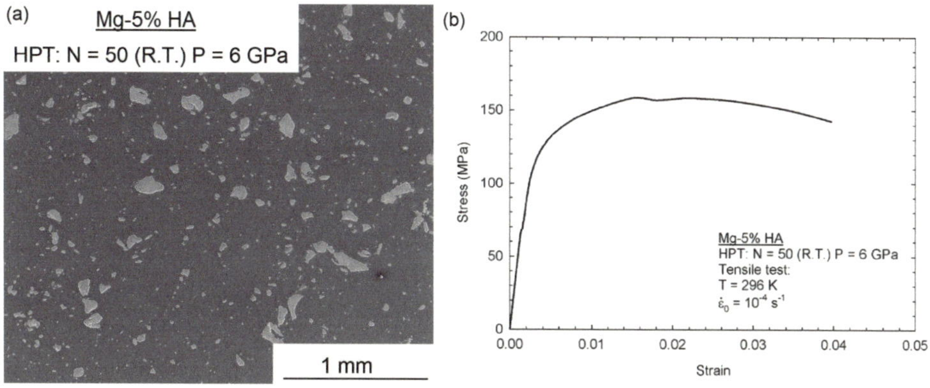

Figure 5.
(a) SEM backscattered electron image of the mid-radius area of the Mg-HA composite, and (b) it respective stress vs. strain curve [50]. MDPI-open access.

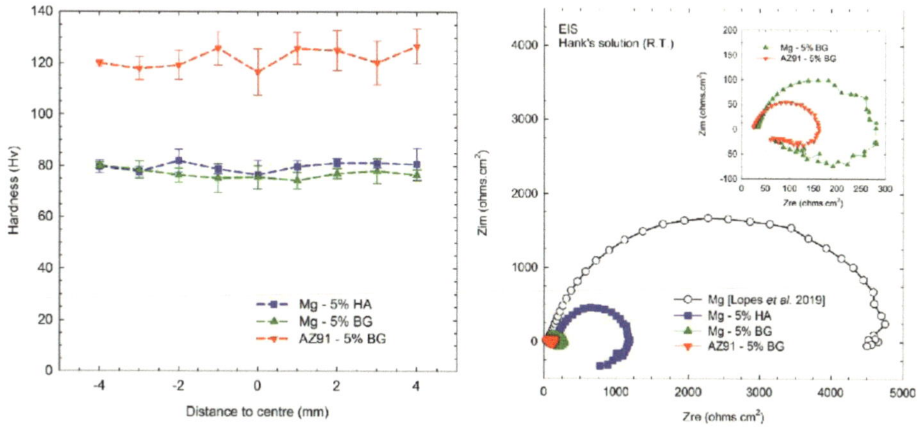

Figure 6.
(a) Microhardness distribution along the disc diameter in the different Mg composites, and (b) electrochemical impedance spectroscopy tests in Hank's solution for the different composites [50]. MDPI-open access.

than for the Mg-5%BG reported by Castro et al. [50] due to the smaller size and morphology of HA particles, what facilitates a better continuity of the matrix and less stress concentration on the Mg-5%HA composite. In relation to pure Mg, immersion tests of the Mg-5%HA composite leads to a resulting superficial layer that appears to be thicker and the corrosion seems to be more generalized, what contributes for the diminishing of the corrosion rate after longer immersion times (**Figure 7**).

In another recent work, a combined method of cyclic extrusion compression (CEC), equal channel angular pressing (ECAP) and conventional extrusion were employed to fabricate Mg-HA composites with 2, 5 and 10 wt% HA at 400°C [54]. Among all samples, the Mg-5HA composite presented the finest grains and exhibited the highest hardness and strength. Hydrogen evolution tests and potentiodynamic polarization tests showed that among all samples, this composite has the lowest values of evolved hydrogen, weight loss and corrosion current density. Its corrosion resistance is also better than in cast and extruded pure Mg. The CECAP-FE process proved to be able to achieve a good and homogenous reinforcement dispersion until 5 wt% of HA. Adding more reinforcement deteriorate mechanical properties and corrosion resistance, what is directedly related to the development of the inhomogeneous microstructure and HA agglomerations.

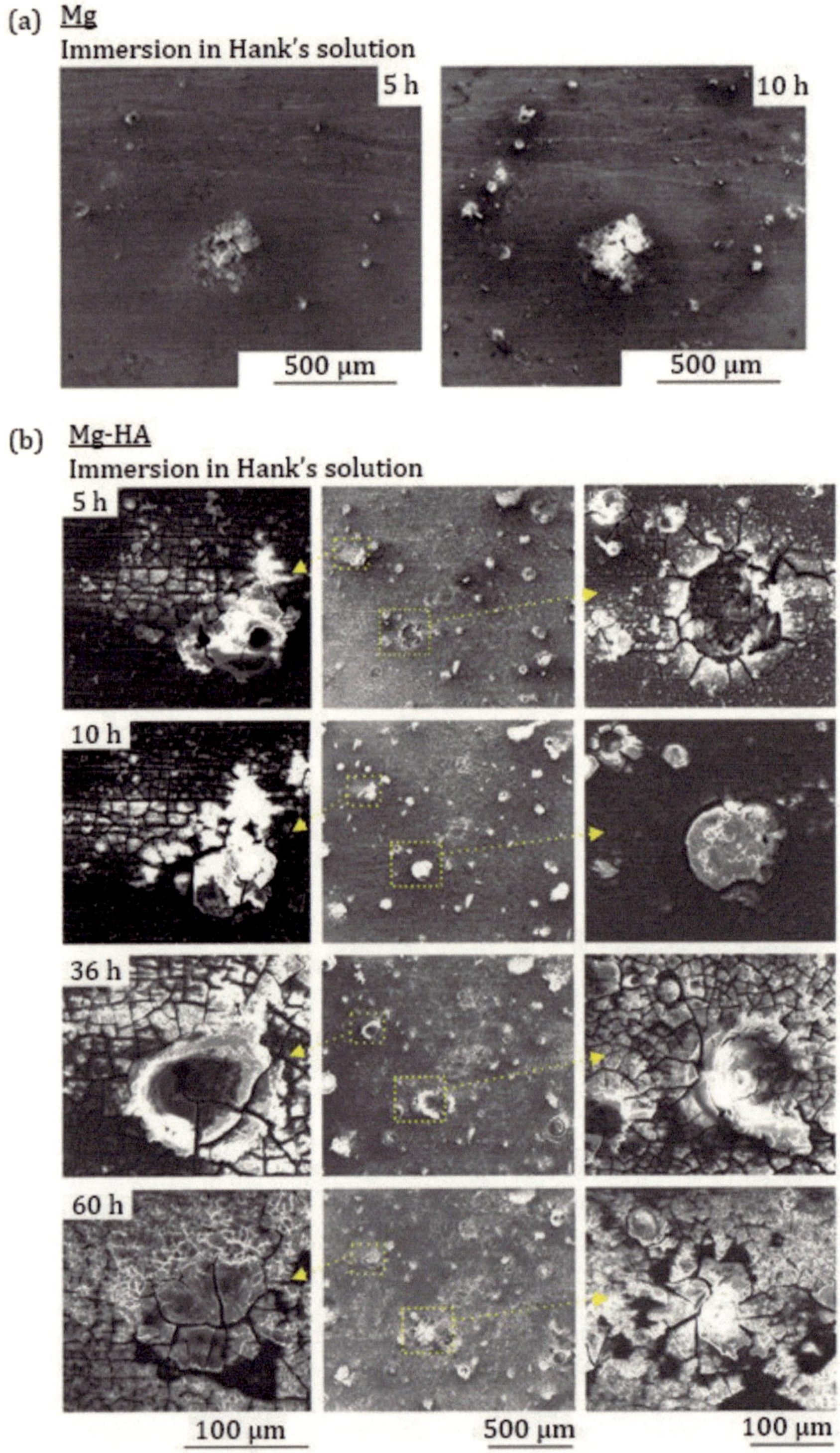

Figure 7.
(a) Mg and (b) Mg-HA surfaces after immersion in Hank's solution. (reproduced with permission [52]. Copyright 2020, AEM/license number: 4924491163715).

The thin surface layer of magnesium oxides and hydroxides formed when these materials are exposed to neutral or alkaline solutions have time-dependent stability and, in many studies conducted on magnesium alloys with ultra-fine grains, the results for the evolution of this stability over time are inconsistent [55]. For example, in the research works of Minárik et. al [56] and Zhang et. al [57], there was an increase in the stability of Mg oxides and hydroxides over time, while Song et. al. [58] observed a decrease in this stability. There are also studies in which growth and decreased stability were reported during the analysis period [59]. This factor is relevant, and as the results of polarization tests or hydrogen evolution - which involve immersion - may be affected by the protective oxide instability [55]. The contradictory conclusions of corrosion studies in pure magnesium and magnesium alloys may be due to these microstructural heterogeneities resulting from the recrystallization dynamics [55, 58–61].

5. Concluding remarks

Magnesium-based composites are recognized as promising materials for biomedical applications due to the combination of good mechanical properties, low density, and possibility to achieve a slower corrosion rate and enhance healing by the presence of bioactive materials.

Several studies conducted by many research groups all around the word leads to the conclusion that a more controllable corrosion rate relies on a homogeneous second phase distribution, and refined matrix absent of cracks or stress concentrators.

The use of fabrication methods based on the plastic deformation of metallic particles requires less time and lower temperatures to produce whole dense materials in comparison to techniques primarily relied on diffusional processes. These processes enable the achievement of smaller grain sizes and thus better mechanical properties and slower corrosion rates. Besides, the use of SPD for processing Mg-based materials does not compromise biocompatibility.

The possibility to create composites without using high-temperature steeps is interesting not only due to the operational and energetical advantages (it is less energy and time consuming, generates no fumes, and avoid grain growth), but also due to the potential of creating controlled drug delivery system with medicines that are temperature sensitive and could deteriorate during conventional processing.

It appears that combining materials and techniques could be more efficient than a singular approach to overcome the current issues to produce an ideal biodegradable implant. Yet finding the best materials combination, adjusting their proportion, and selecting the most suitable fabrication route with optimized parameters is still challenging.

Acknowledgements

The authors are grateful for the publisher support.

Conflict of interest

The authors declare no conflict of interest.

Author details

Moara Marques de Castro[1]*, Débora Ribeiro Lopes[2] and Leonardo Viana Dias[3]

1 Department of Metallurgical and Materials Engineering, Universidade Federal de Minas Gerais, Belo Horizonte, Brazil

2 Department of Chemical Engineering, Universidade Federal de Minas Gerais, Belo Horizonte, Brazil

3 Department of Metallurgical Engineering, Instituto Federal de Minas Gerais, Ouro Branco, Brazil

*Address all correspondence to: moara.m.c@gmail.com

References

[1] Kuśnierczyk, K. and M. Basista, *Recent advances in research on magnesium alloys and magnesium–calcium phosphate composites as biodegradable implant materials.* Journal of Biomaterials Applications, 2016.**31**(6): p. 878-900.

[2] Brar, H.S., et al., *Magnesium as a biodegradable and bioabsorbable material for medical implants.* JOM, 2009. **61**(9): p. 31-34.

[3] Kamrani, S. and C. Fleck, *Biodegradable magnesium alloys as temporary orthopaedic implants: a review.* BioMetals, 2019.

[4] Ali, M., M.A. Hussein, and N. Al-Aqeeli, *Magnesium-based composites and alloys for medical applications: A review of mechanical and corrosion properties.* Journal of Alloys and Compounds, 2019. **792**: p. 1162-1190.

[5] Katarivas Levy, G., J. Goldman, and E. Aghion, *The Prospects of Zinc as a Structural Material for Biodegradable Implants—A Review Paper.* Metals, 2017. 7(10).

[6] Davis, J.R., *Handbook of materials for medical devices*. 2003.

[7] Bommala, V.K., M.G. Krishna, and C.T. Rao, *Magnesium matrix composites for biomedical applications: A review.* Journal of Magnesium and Alloys, 2019. 7(1): p. 72-79.

[8] Barrère, F., et al., *Advanced biomaterials for skeletal tissue regeneration: Instructive and smart functions.* Materials Science and Engineering: R: Reports, 2008. **59**(1): p. 38-71.

[9] Niinomi, M. and M. Nakai, *Titanium-Based Biomaterials for Preventing Stress Shielding between Implant Devices and Bone.* International Journal of Biomaterials, 2011. **2011**: p. 836587.

[10] Mi, Z.R., et al., *Problem of Stress Shielding and Improvement to the Hip Implat Designs: A Review.* Journal of Medical Sciences, 2007. 7: p. 460-467.

[11] Pogorielov, M., et al., *Magnesium-based biodegradable alloys: Degradation, application, and alloying elements.* Interventional Medicine & Applied Science, 2017. **9**(1): p. 27-38.

[12] Bowen, P.K. and W.H. Sillekens, *Overview: Magnesium-Based Biodegradable Implants.* JOM, 2016. **68**(4): p. 1175-1176.

[13] Gröber, U., J. Schmidt, and K. Kisters, *Magnesium in prevention and therapy.* Nutrients, 2015. 7(9): p. 8199-8226.

[14] Al Alawi, A.M., S.W. Majoni, and H. Falhammar, *Magnesium and human health: perspectives and research directions.* International Journal of Endocrinology, 2018. **2018**.

[15] Atrens, A., et al., *Review of recent developments in the field of magnesium corrosion.* Advanced Engineering Materials, 2015. **17**(4): p. 400-453.

[16] Radha, R. and D. Sreekanth, *Insight of magnesium alloys and composites for orthopedic implant applications – a review.* Journal of Magnesium and Alloys, 2017. **5**(3): p. 286-312.

[17] Thomas, S., et al., *Corrosion mechanism and hydrogen evolution on Mg.* Current Opinion in Solid State and Materials Science, 2015. **19**(2): p. 85-94.

[18] Zhang, C.Z., et al., *The microstructure and corrosion resistance of biological Mg-Zn-Ca alloy processed by high-pressure torsion and subsequently annealing.* Journal of Materials Research, 2017. **32**(6): p. 1061-1072.

[19] Birbilis, N., et al., *Grain character influences on corrosion of ECAPed pure magnesium.* Corrosion Engineering,

Science and Technology, 2010. **45**(3): p. 224-230.

[20] Song, G. and A. Atrens, *Understanding magnesium corrosion—a framework for improved alloy performance.* Advanced engineering materials, 2003. **5**(12): p. 837-858.

[21] Maradze, D., et al., *High Magnesium Corrosion Rate has an Effect on Osteoclast and Mesenchymal Stem Cell Role During Bone Remodelling.* Scientific Reports, 2018. **8**(1): p. 10003.

[22] Xin, R., et al., *Texture effect on corrosion behavior of AZ31 Mg alloy in simulated physiological environment.* Materials letters, 2012. **72**: p. 1-4.

[23] Saha, P., et al., *Effects of grain refinement on the biocorrosion and in vitro bioactivity of magnesium.* Materials Science and Engineering: C, 2015. **57**: p. 294-303.

[24] Argade, G.R., S.K. Panigrahi, and R.S. Mishra, *Effects of grain size on the corrosion resistance of wrought magnesium alloys containing neodymium.* Corrosion Science, 2012. **58**: p. 145-151.

[25] Silva, C.L.P., et al., *Effect of severe plastic deformation on the biocompatibility and corrosion rate of pure magnesium.* Journal of Materials Science, 2017. **52**(10): p. 5992-6003.

[26] Chen, J., et al., *Mechanical properties of magnesium alloys for medical application: A review.* Journal of the Mechanical Behavior of Biomedical Materials, 2018. **87**: p. 68-79.

[27] Persaud-Sharma, D. and A. McGoron, *Biodegradable Magnesium Alloys: A Review of Material Development and Applications.* Journal of biomimetics, biomaterials, and tissue engineering, 2012. **12**: p. 25-39.

[28] Kirkland, N.T., et al., *Performance-driven design of Biocompatible Mg alloys.* JOM, 2011. **63**(6): p. 28-34.

[29] Haghshenas, M., *Mechanical characteristics of biodegradable magnesium matrix composites: A review.* Journal of Magnesium and Alloys, 2017. **5**(2): p. 189-201.

[30] Glenske, K., et al., *Applications of metals for bone regeneration.* International journal of molecular sciences, 2018. **19**(3): p. 826.

[31] Xynos, I.D., et al., *Ionic Products of Bioactive Glass Dissolution Increase Proliferation of Human Osteoblasts and Induce Insulin-like Growth Factor II mRNA Expression and Protein Synthesis.* Biochemical and Biophysical Research Communications, 2000. **276**(2): p. 461-465.

[32] Wan, Y., et al., *Mechanical and biological properties of bioglass/magnesium composites prepared via microwave sintering route.* Materials & Design, 2016. **99**: p. 521-527.

[33] Kiani, F., C. Wen, and Y. Li, *Prospects and strategies for magnesium alloys as biodegradable implants from crystalline to bulk metallic glasses and composites—A review.* Acta Biomaterialia, 2020. **103**: p. 1-23.

[34] Jaiswal, S., A. Dubey, and D. Lahiri, *In Vitro Biodegradation and Biocompatibility of Mg–HA-Based Composites for Orthopaedic Applications: A Review.* Journal of the Indian Institute of Science, 2019. **99**(3): p. 303-327.

[35] Jingxin, Y., et al., *A review on the exploitation of biodegradable magnesium-based composites for medical applications.* Biomedical Materials, 2018. **13**(2): p. 022001.

[36] Liu, Q., et al., *The influence of particles size and its distribution on the degree of stress concentration in particulate reinforced metal matrix composites.* Materials Science and Engineering: A, 2018. **731**: p. 351-359.

[37] Gao, J.H., et al., *Homogeneous corrosion of high pressure torsion treated Mg–Zn–Ca alloy in simulated body fluid.* Materials Letters, 2011. **65**(4): p. 691-693.

[38] Kumar Khanra, A., et al., *Microstructure and mechanical properties of Mg-HAP composites*. Vol. 33. 2010, Bulletin of Materials Science. 43-47.

[39] Khanra, A.K., et al., *Comparative property study on extruded Mg–HAP and ZM61–HAP composites.* Materials Science and Engineering: A, 2010. **527**(23): p. 6283-6288.

[40] Ye, X., et al., *In vitro corrosion resistance and cytocompatibility of nano-hydroxyapatite reinforced Mg–Zn–Zr composites.* Journal of Materials Science: Materials in Medicine, 2010. **21**(4): p. 1321-1328.

[41] Sun, J.e., et al., *The effect of nano-hydroxyapatite on the microstructure and properties of Mg–3Zn–0.5 Zr alloy.* Journal of Composite Materials, 2014. **48**(7): p. 825-834.

[42] Kumar, S., et al., *Enhanced mechanical properties and hydrophilic behavior of magnesium oxide added hydroxyapatite nanocomposite: A bone substitute material for load bearing applications.* Ceramics International, 2020. **46**(10, Part B): p. 16235-16248.

[43] Stüpp, C.A., et al. *Powder Metallurgical Synthesis of Biodegradable Mg-Hydroxyapatite Composites for Biomedical Applications*. Trans Tech Publ.

[44] Gu, X., et al., *Microstructure, mechanical property, bio-corrosion and cytotoxicity evaluations of Mg/HA composites.* Materials Science and Engineering: C, 2010. **30**(6): p. 827-832.

[45] Jaiswal, S., et al., *Mechanical, corrosion and biocompatibility behaviour of Mg-3Zn-HA biodegradable composites for orthopaedic fixture accessories.* Journal of the Mechanical Behavior of Biomedical Materials, 2018. **78**: p. 442-454.

[46] Setyadi, I., et al., *Microstructure and microhardness of carbonate apatite particle-reinforced Mg composite consolidated by warm compaction for biodegradable implant application.* Materials Research Express, 2020.

[47] Salleh, E.M., et al., *In Vitro Biodegradation and Mechanical Properties of Mg-Zn Alloy and Mg-Zn-Hydroxyapatite Composite Produced by Mechanical Alloying for Potential Application in Bone Repair.* Metallurgical and Materials Transactions A, 2018. **49**(11): p. 5888-5903.

[48] Ratna Sunil, B., et al., *Processing and mechanical behavior of lamellar structured degradable magnesium–hydroxyapatite implants.* Journal of the Mechanical Behavior of Biomedical Materials, 2014. **40**: p. 178-189.

[49] Xiong, G., et al., *Characterization of biomedical hydroxyapatite/magnesium composites prepared by powder metallurgy assisted with microwave sintering.* Current Applied Physics, 2016. **16**(8): p. 830-836.

[50] Castro, M.M., et al., *Magnesium-Based Bioactive Composites Processed at Room Temperature.* Materials (Basel, Switzerland), 2019. **12**(16): p. 2609.

[51] Dobatkin, S.V., et al., *Strength, corrosion resistance, and biocompatibility of ultrafine-grained Mg alloys after different modes of severe plastic deformation.* IOP Conference Series: Materials Science and Engineering, 2017. **194**: p. 012004.

[52] Lopes, D., et al., *Corrosion Behavior in Hank's Solution of a Magnesium–Hydroxyapatite Composite Processed by High-Pressure Torsion.* Advanced Engineering Materials, 2020. **n/a**(n/a): p. 2000765.

[53] Castro, M.M., et al., *Consolidation of magnesium and magnesium-quasicrystal composites through high-pressure torsion.* Letters on Materials, 2019. **9**(4): p. 546-550.

[54] Torabi, H., et al., *Microstructure, mechanical properties and bio-corrosion properties of Mg-HA bionanocomposites fabricated by a novel severe plastic deformation process.* Ceramics International, 2020. **46**(3): p. 2836-2844.

[55] Miyamoto, H., *Corrosion of ultrafine grained materials by severe plastic deformation, an overview.* Materials Transactions, 2016. **57**(5): p. 559-572.

[56] Minárik, P., R. Král, and B. Hadzima, Substantially higher corrosion resistance in AE42 magnesium alloy through corrosion layer stabilization by ECAP treatment. Acta Physica Polonica-Series A General Physics, 2012. 122(3): p. 614.

[57] Zhang, F., et al., Enhanced biodegradation behavior of ultrafine-grained ZE41A magnesium alloy in Hank's solution. Progress in Natural Science: Materials International, 2013. 23(4): p. 420-424.

[58] Song, D., et al., Corrosion behaviour of bulk ultra-fine grained AZ91D magnesium alloy fabricated by equal-channel angular pressing. Corrosion Science, 2011. 53(1): p. 362-373.

[59] Vrátná, J., et al., Room temperature corrosion properties of AZ31 magnesium alloy processed by extrusion and equal channel angular pressing. Journal of Materials Science, 2013. 48(13): p. 4510-4516.

[60] Song, D., et al., Corrosion behavior of equal-channel-angular-pressed pure magnesium in NaCl aqueous solution. Corrosion Science, 2010. 52(2): p. 481-490.

[61] Hamu, G.B., D. Eliezer, and L. Wagner, The relation between severe plastic deformation microstructure and corrosion behavior of AZ31 magnesium alloy. Journal of alloys and compounds, 2009. 468(1-2): p. 222-229.

Chapter 6

Solidification Processing of Magnesium Based In-Situ Metal Matrix Composites by Precursor Approach

Nagaraj Chelliah Machavallavan, Rishi Raj and M.K. Surappa

Abstract

In-situ magnesium based metal matrix composites (MMCs) belong to the category of advanced light weight metallic composites by which ceramic dispersoids are produced by a chemical reaction within the metal matrix itself. In-situ MMCs comprised uniform distribution of thermodynamically stable ceramic dispersoids, clean and unoxidized ceramic-metal interfaces having high interfacial strength. In last two decades, investigators have been collaborating to explore the possibility of enhancing the high temperature creep resistance performance in polymer-derived metal matrix composites (P-MMCs) by utilizing polymer precursor approach. A unique feature of the P-MMC process is that since all constituents of the ceramic phase are built into the polymer molecules itself, there is no need for a separate chemical reaction between the host metal and polymer precursor in order to form in-situ ceramic particles within the molten metal. Among the different polymer precursors commercially available in the market, the silicon-based polymers convert into the ceramic phase in the temperature range of 800–1000°C. Therefore, these Si-based polymers can be infused into molten Mg or Mg-alloys easily by simple stir-casting method. This chapter mainly focuses on understanding the structure–property correlation in both the Mg-based and Mg-alloy based in-situ P-MMCs fabricated by solidification processing via polymer precursor approach.

Keywords: in-situ metal matrix composites, magnesium, Mg-alloys, solidification, polymer derived ceramics

1. Introduction

In-situ metal matrix composites offer superior microstructural/mechanical characteristics compared to their conventional counterparts. In-situ MMCs are characterized by uniform dispersion of fine-sized thermodynamically stable ceramic particles, clean and unoxidized ceramic-metal interfaces having high interfacial strength, improved hardness/yield strength and elastic modulus [1]; In-situ composite fabrication techniques are gaining attraction among the researchers since this techniques overcome several processing related issues such as non-uniformity of particle distribution, poor wettability and weak interfaces associated with composites produced by conventional processing techniques [2–3]. In-situ processing methods use a chemical reaction between liquid–gas, liquid–liquid, liquid–solid, and mixed salts to generate fine ceramic particles in the metal matrix. For instance, silicon, dissolved in copper, may be converted into a dispersion of silica particles by reacting with oxygen diffusing in from the environment [4]. Enormous data are available in the composite literature regarding the processing of Al-based in-situ composites whereas data on in-situ Mg-based composites seems to be very limited [5–6].

In last two decades, investigators have been collaborating to enhance the high temperature creep resistance of metal matrix composites by utilizing polymer precursor approach [7–19]. A noteworthy feature of this polymer precursor approach is that no chemical reaction between polymer precursor and the host Mg is required to produce in-situ ceramic particles because the polymer contains the constituents of ceramic phases within organic molecules itself [14–15]. The pyrolysis is a highly reactive process, accompanied by the evolution of hydrogen, which disperses the ceramic phase into nanoscale or sub-micron constituents [1]. However, the two critical main issues were identified in earlier work [14–15] as follows: (i) the chemical reaction between polymer precursor and magnesium melt results in the formation of brittle Mg_2Si particles at pyrolysis temperature ranging from 800 to 1000°C. These brittle Mg_2Si ceramic phase significantly reduce the ductility of fabricated composite. In addition, formation of Mg_2Si phase leads to reduction in the amount of polymer precursor available for in-situ generation of sub-micron or nano-sized SiCNO particles, and (ii) most of the PDC particles are pushed away by the solidification front and get segregated at the grain boundaries. Such grain boundary segregation limits any enhancement in the mechanical properties of the final in-situ MMCs. These two critical issues can be mitigated by adopting the following research methodologies [13] namely (i) The intensity of Mg_2Si formation can be minimized by lowering the pyrolysis temperature from 800 to 700°C. (ii) The issue of grain boundary segregation can be reduced by adopting friction stir processing (FSP) treatment. The severe plastic deformation within the nugget zone is expected to create fragmentation in the larger SiCNO particles leading to nano-sized particles, and stirring action arising during FSP tool aids in achieving the uniform dispersion of particles throughout the Mg matrix [1].

The microstructural design of light weight MMCs for creep resistant applications involves the uniform dispersion of thermally stable nano-scale ceramic particles throughout the grain matrix [1]. The potential for incorporating nano-sized PDC particles into a light weight metal is made possible by polymer precursor approach [1]. The direct injection of liquid polysilazane precursor into molten metal and the subsequent casting process seems to be most advantageous and viable route of producing P-MMCs. If nano-scale particles of PDCs are dispersed uniformly in the magnesium matrix, it can be expected to impede dislocation motion and resist coarsening at high temperatures, thereby holding the promise of a new genre of light weight polymer-derived P-MMCs [1]. Therefore, the present chapter mainly focuses on understanding the correlation between structure-properties of in-situ magnesium matrix composites containing SiCNO particles via polymer precursor approach.

2. In-situ polymer precursor approach

Rishi Raj and M. K. Surappa jointly [14] patented a novel pyrolysis infiltration process (PIP) to fabricate Mg-based P-MMCs using liquid metallurgical route. The PIP process has great generality because different organic precursors for producing oxide, carbides, nitrides, and borides are commercially available as indicated in **Table 1** [1].

Polymer precursors	Ceramic phases	Merits	Demerits
Polysilazanes	Si-C-N-O (amorphous)	Commercially available Mostly air tolerant Can be thermally cross–linked High molecular weight Environmentally benign Pyrolyzes at 700–1000°C	Can form silicides
Siloxanes	Si-C-O (amorphous)		
Carbosilanes	Si-C (crystalline)		
Alkoxides	Metal oxides, e.g. SiO_2, ZrO_2, HfO_2, Al_2O_3 etc.	Commercially available Moderately air tolerant. Pyrolyzes at 400°C	Cannot be cross–linked High vapor pressure

Table 1.
Types of organic precursors that can serve as a source for ceramics in MMCs [1, 15].

Also, the process would permit the addition of the desired volume fractions of the ceramic enabling the nanostructural design, and production of P-MMCs with a wide range of mechanical properties, meant especially for high temperature creep resistant applications [1]. When the melting point of a metal lies below 800°C, the Si-based polymer can be added into the molten metal by a stir-casting method. Magnesium and aluminum matrix composites are typical examples for this PIP approach. The conversion of the polymer into the ceramic phase releases hydrogen and small amounts of hydrocarbons. During in-situ processing these gaseous species must migrate to the free surface in a way to avoid any porosity in the final castings. This can be an issue in aluminum composite castings; although it does not appear to constrain the production of magnesium composite castings. Originally, the standard procedure recommended by inventor of this PIP process [14] is to add the cross-linked polymer powder just above the melting point and then superheat the metal to 800°C to convert the organic into the ceramic phase. Later, they realized that direct injection of the liquid polymer is even most simpler [15], and holds greater potential for futuristic development of polymer derived metal matrix composites (P-MMCs).

Figure 1 depicts the schematic diagram of casting setup used to fabricate in-situ Mg/Mg-alloys based P-MMCs [1]. The systematic procedures for composite fabrication can be described in the following steps; [1].

i. Firstly, one kilogram of Mg blocks (Pure Mg or Mg-alloys (AZ91 and AE44 series) was melted in a steel crucible using an electrical resistance furnace at a temperature range of 700–900°C.

ii. In order to eliminate the flammability and risk of fire hazards with molten Mg, the steel crucible was then continuously purged with Ar-5%SF6 gas mixture.

iii. The molten Mg was degassed by argon (99.999% purity) gas for the period of about 10 minutes before reinforcement. The composites are reinforced with a projected volume fraction of 2.5% of PDC particles during stir-casting process.

iv. The melt was mechanically stirred by a 3-axial stirrer blade at 600 rpm to create a good vortex in the melt. Subsequently, required amount of liquid polymer or cross-linked polymer (pre-heated at a temperature of 200°C) was injected into the melt and the stirring was continued for next 15 minutes in order to ensure completion of in-situ pyrolysis.

v. The polymer to ceramic conversion of cross-linked polymer occurs in single stage (pyrolysis) whereas the liquid polymer conversion takes place in two stages (both cross-linking as well as pyrolysis).

vi. In the temperature range of 700–800°C, the bonds between carbon and hydrogen in polysilazane precursor become unstable and hydrogen gas is released. These dangling carbon atoms self-assemble into graphene-like network or retained as free carbon clusters [20].

vii. After the completion of polymer-to-ceramic conversion, the amorphous ceramic phase is constituted by a mixture of various covalent bonds such as Si-C, Si-O, Si-N and C-C [20]. As the final PDC contains some of residual oxygen, the general chemical composition of PDC can be expressed as SiCNO.

viii. After processing, the mixed molten slurry was bottom-poured into a pre-heated split-molds made of steel.

The designation of these fabricated Mg matrix composites has been indexed as PP700, PP800, PL700, and PL800. Mg-alloys based P-MMCs were designated as PP900-AZ, and PP900-AE which were fabricated using AZ91 and AE44 Mg-alloys as matrix materials, respectively. Here, PP refers to polymer derived composites made using cross-linked powder, and PL indicates polymer derived composites made using as-received liquid ceraset. The last 3 digits refer to process or pyrolysis temperature. In addition, the castings from pure magnesium, and unreinforced Mg-alloys were also fabricated using similar processing parameters.

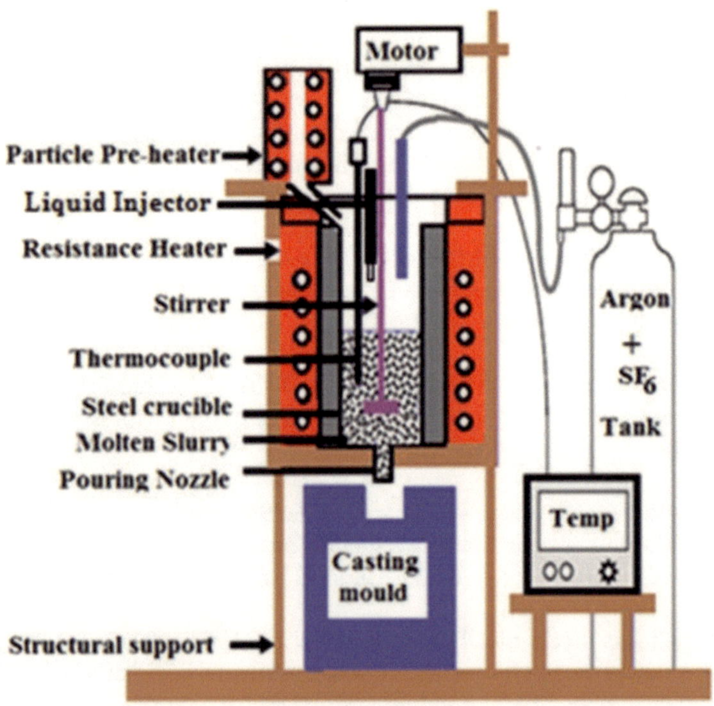

Figure 1.
Schematic diagram of stir-casting furnace used for fabricating in-situ composites [1].

3. Structure-property correlation of as-cast In-situ Mg-based metal matrix composites

3.1 Microstructural characteristics

Figure 2 shows the microstructural evolution of as-cast magnesium composites fabricated under four different processing conditions. It is clearly evident from microstructural analysis that the as-cast PP800/PL800 specimen exhibit bi-modal grain size distribution, whereas the as-cast PP700/PL700 specimen represents more or less uni-modal grain size distribution. Such difference in microstructural characteristics arises because of the fact that higher amount of heterogeneous nucleation sites is available for producing Mg crystals in PP800/PL800 specimen as they contain both the SiCNO and Mg2Si particles. However, it is mostly SiCNO particles with negligible amount of Mg2Si particles are observed for the case of PP700/PL700 specimen. This could be also easily inferred from XRD spectra (**Figure 3**). For instance, the intensity of Mg_2Si peaks appears to be stronger in composites processed at temperature of 800°C and weak in composites processed at 700°C. Notice the fact that since SiCNO ceramic phases are amorphous in nature, corresponding peaks have not appeared in diffraction spectra. Most importantly, it is also observed that the peaks of SiO_2 and $Mg(SiO_4)$ phases are apparent only in the as-cast PL700/PP700 specimen, however, but such peaks are absent in as-cast PL800/PP800 specimen. It is likely expected that because of lower solubility of Si-atoms in the molten magnesium, the chemical reaction takes place between the magnesium and silicon to form Mg_2Si crystal according to the following equation [21]:

$$2Mg(l) + Si(s) \rightarrow Mg_2Si(s) \tag{1}$$

According to the chemical reaction (1), it can be estimated that the change in Gibbs free energy values (ΔG_f) are found to be −63.578 kJ and – 57.926 kJ for the

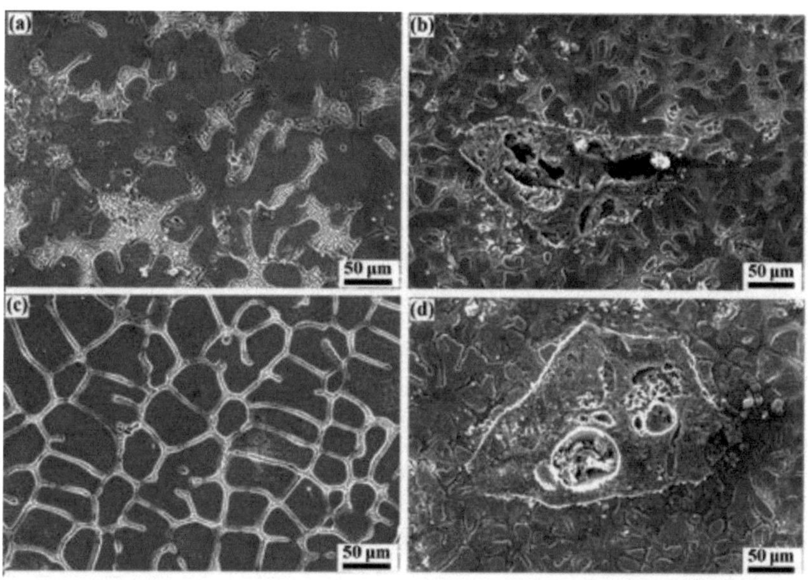

Figure 2.
Microstructural evolution of in-situ Mg matrix composites in as-cast condition (a) PP700 composite (b) PP800 composite (c) PL700 composite and (d) PL800 composite [1].

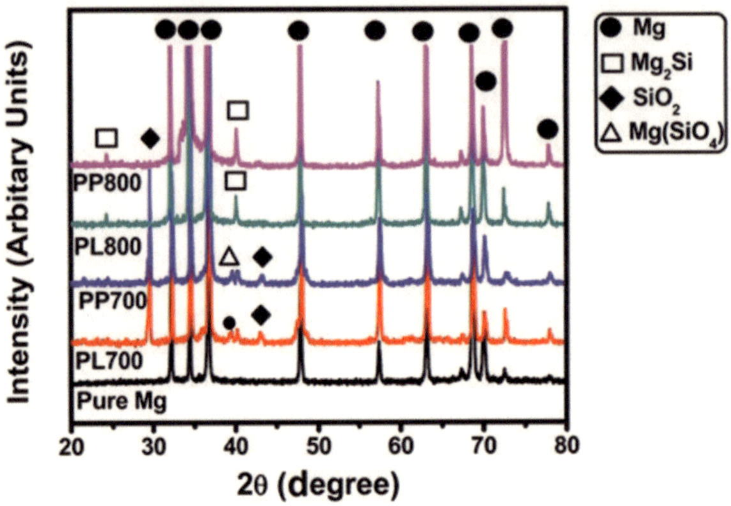

Figure 3.
XRD spectra of pure Mg and in-situ Mg matrix composites [1].

processing temperatures of 800°C and 700°C, respectively. The higher negative value at ΔG_f at 800°C represents that the tendency for Mg_2Si formation is increased by increasing the pyrolysis temperature from 700 to 800°C. In other words, it is possible to minimize the formation of Mg_2Si phase by reducing the processing temperature from 800 to 700°C during in-situ pyrolysis of the polymer precursor.

Inem et al. [22] reported that there is no extensive direct chemical reaction occurs between Mg and SiC particle to form Mg_2Si crystal at 900°C in the SiCp particles reinforced AZ91 Mg-alloy. However, they predicted that SiO_2 scale that forms on SiC particle can react with molten Mg to form Mg_2Si according to the following chemical reaction;

$$4\,Mg(l) + SiO_2(s) \rightarrow 2MgO(s) + Mg_2Si(s) \tag{2}$$

This data indicates that some amount of SiO_2 scale must be always present in order to produce any Mg_2Si phase in the composite. As it can be seen in **Figure 3**, XRD data shows the presence of SiO_2 peaks only in the PL700 specimen but not in the as-cast PL800 specimen. This means that the PL800 specimen consumes SiO_2 phase completely to form Mg_2Si crystal whereas some free SiO_2 is left behind in the PL700 specimen due to lack of formation of Mg_2Si particles.

According to constitutional supercooling theory, the ratio of temperature gradient (G) to growth rate (R) determines the grain morphology of the final castings during solidification [23]. This theory predicts that the microstructures can be changed from cellular to columnar dendritic, and then to equiaxed dendritic morphology if the solidification condition possess low G/R ratio [23]. It is worthwhile to mention the fact that constitutional supercooling theory can also be applied to the solidification of metal matrix composites if the tip of the solidification front contains certain level of solute impurities. Kim et al. [24] pointed out that any change in the direction of heat flux resulting in different microstructures during solidification of the metal matrix composites. They explained that if the direction of heat flux is same to that of crystal growth, then equiaxial dendritic growth occurs under solidification condition. On the contrary, columnar growth occurs in the final castings if the directions of heat flux and crystal growth are different or antiparallel. The second scenario is quiet possible if heat transfer occurs by conduction mechanism across the mold material. This means that positive temperature gradient exists in front of the solidification front for columnar microstructures and negative temperature gradient for the case of dendritic grain microstructures. We believed that the presence of Mg_2Si particles at 800°C in the molten magnesium act as an effective heterogeneous nucleation site to produce equiaxial dendritic grain simultaneously throughout the molten magnesium. However, any absence of Mg_2Si particles in the molten magnesium fabricated at 700°C leads to produce columnar grain from the edge to center of the mold. Therefore, the microstructures of the final in-situ magnesium composites should be different as they were processed under two different solidification conditions. As shown in **Figure 2**, the microstructures of PP700/PL700 composites show more or less columnar grain morphology while the grain morphology of PP800/PL800 composites appears to be predominantly dendritic in nature. From experimental observation, it can be understood clearly that the grain morphology in the final castings can be controlled by imposing variation in G/R ratio and heat flux direction by changing the process or pyrolysis temperature of the in-situ magnesium composites.

Figure 4 shows the microstructures of as-cast composites fabricated under different processing conditions along with corresponding energy-dispersive spectroscopy (EDS) data. Irrespective of the processing condition, all the EDS spectra show a significant presence of Si, C, N and O. Notice that nominal oxygen content in the SiCNO ceramic particles of in-situ composites fabricated at 800°C is significantly higher than that of composites processed at 700°C [1]. Such difference in

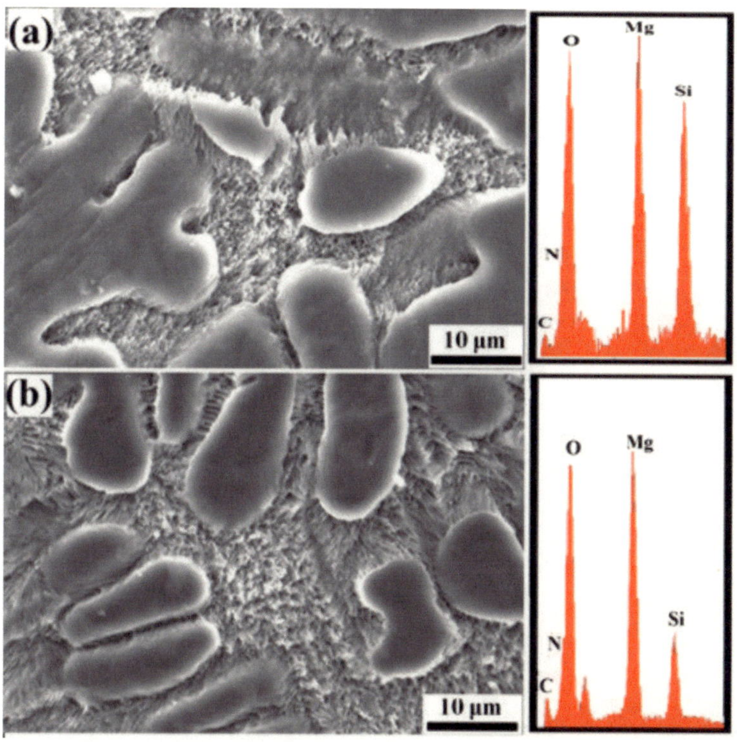

Figure 4.
SEM micrographs of in-situ Mg matrix composites in as-cast condition along with corresponding EDS spectra (a) PP800 composite and (b) PL800 composite [1].

oxygen content arises because of increase in the partial oxidation of organic phase while it was being added into the magnesium melt [25]. Microstructural characterization reveals the fact that the intensity of grain boundary segregation of PDC particles appears to be significantly larger in PP800/PL800 composites as compared to that of PP700/PL700 composites as seen in **Figure 5** [1]. As long as the ceramic particles are being pushed away by the solidification front, the composites having dendritic grain morphology are expected to hold more number of PDC particles at the grain boundaries as compared to the composites with columnar grain morphology during solidification. This can be associated with the fact that each dendritic grain is comprised of six orthogonal primary trunks whereas the columnar grain possesses no such primary trunks [21]. Moreover, the particle sizes of PDC ceramic phase vary with processing temperature. For instance, PP700/PL700 composites are characterized by coarsened round shaped particle having size range of 500 to 700 nm while platelet-like shaped particles have a width of 300 to 400 nm and a length of 2 to 4 μm. However, PP800/PL800 composites depicts slightly fine-sized round shaped particle (having a mean size of about 400 to 500 nm) and platelet-like shaped particles have a width of 100 to 200 nm and a length of 1 to 2 μm. The difference in particle sizes of the fabricated composites could be associated with variation in the intensity of fragmentation of PDC particles during in-situ pyrolysis under different processing temperatures [1].

3.2 Mechanical properties

Figure 6 shows the Vickers hardness data of as-cast in-situ Mg-based P-MMCs along with commercial pure Mg. It was found that the value of hardness of as-cast P-MMCs is enhanced by about 40–75% as compared to that of the base metal. Such enhancement is caused by the reinforcement of PDC particles, and grain size reduction of the as-cast composites. Notice that all hardness measurements are performed in the grain matrix or matrix rich regime of the composites, away from

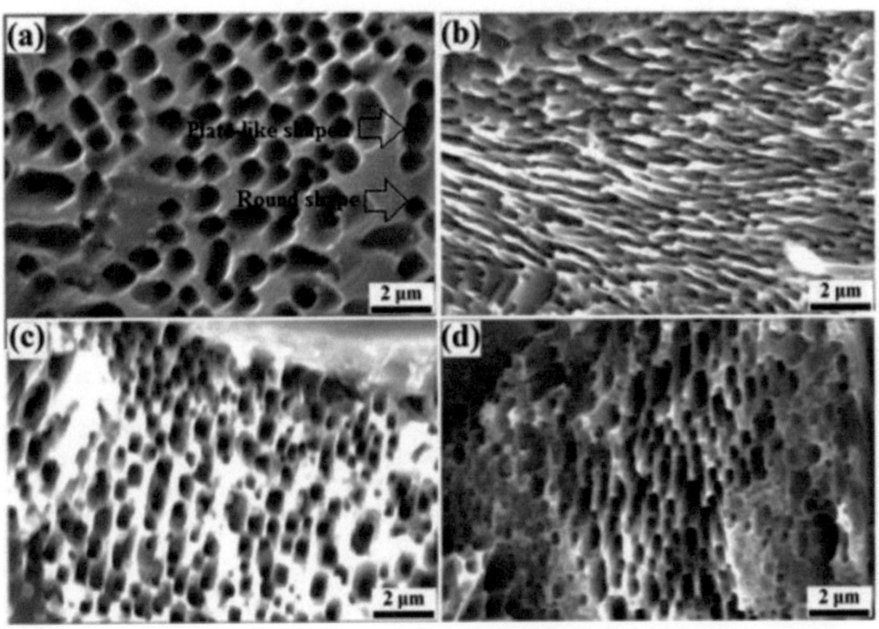

Figure 5.
Particle morphologies of in-situ Mg matrix composites in as-cast condition (a) PP700 composite (b) PP800 composite (c) PL700 composite and (d) PL800 composite [1].

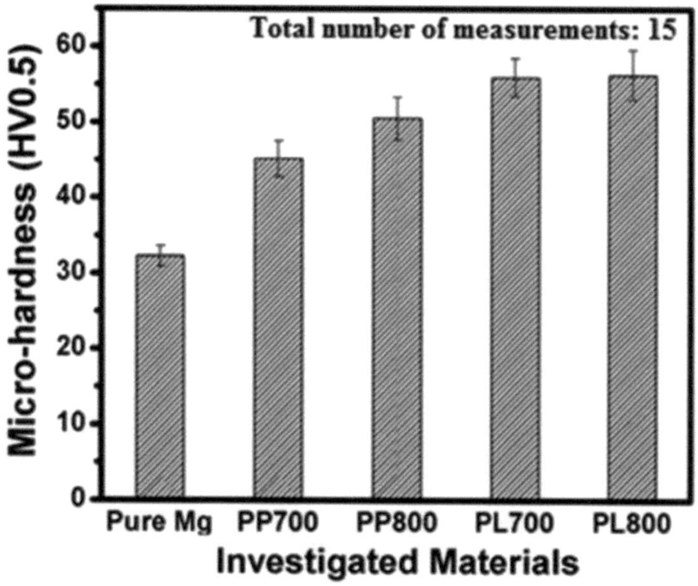

Figure 6.
Micro-hardness properties of pure Mg and in-situ Mg matrix composites [1].

the particles at the grain boundaries but taking care to avoid indentation near the micro-pore area in the as-cast composites in a way to minimize the influence of porosity on the hardness of composites [1]. The hardness data reveals considerable scatter in the values of micro-hardness varying from 45 ± 2 to 56 ± 2 HV. Such a variation arises because of the differences in grain size, volume fraction of in-situ formed PDC particles, and the intensity of grain boundary segregation and formation of Mg_2Si phases in the composites fabricated under different processing conditions [1]. Furthermore, the presence of Mg_2Si crystals in composites fabricated at 800°C exhibits high hardness as compared to that of the composites fabricated at 700°C. This can be attributed to Taylor strengthening mechanism due to thermal mismatch between the matrix and in-situ formed Mg_2Si particles leading to produce significant amount of dislocations at the particle/matrix interface [1].

Figure 7 shows the typical compression true stress-true strain curves of the as-cast composites fabricated under different processing conditions. As depicted in **Figure 7**, the values of compressive yield stress (CYS) and ultimate compression stress (UCS) of all the as-cast composites are significantly higher when compared to pure magnesium [1]. For instance, it can be noticed that the yield stress for the PP800/PL800 specimen is in the range of 77–90 MPa, whereas 85–88 MPa for the PP800/PP700 specimen, but it is only 60 MPa for pure Mg. Moreover, the values of UCS of PL700/PP800 specimen is in the range of 235–237 MPa, enhanced by 10% and 82% as compared to that of PL800 specimen and pure Mg, respectively [1]. Further, the value of strain to failure of PP800/PL800 specimen experiences the lowest strain to failure (9–10%) when compared to that of PL700 specimen (16–18%) and pure Mg (20–22%). The lowest ductility can be closely associated with the presence of brittle Mg_2Si ceramic phase which often impairs the room temperature plasticity of the composites fabricated at 800°C [1].

Figure 8 illustrates the morphology of the fractured surface for the as-cast PL700 and PL800 composites after compression. As shown in **Figure 8**, while the fracture surface of PL700 composite shows the mixed mode of failure including both ductile and cleavage patterns, PL800 composite exhibits cleavage mode

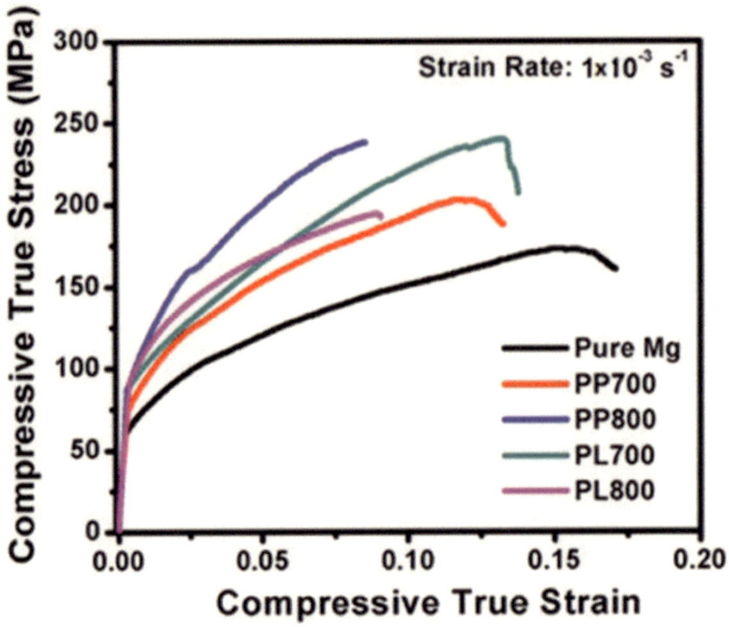

Figure 7.
Compression behavior of pure Mg and in-situ Mg matrix composites at room temperature [1].

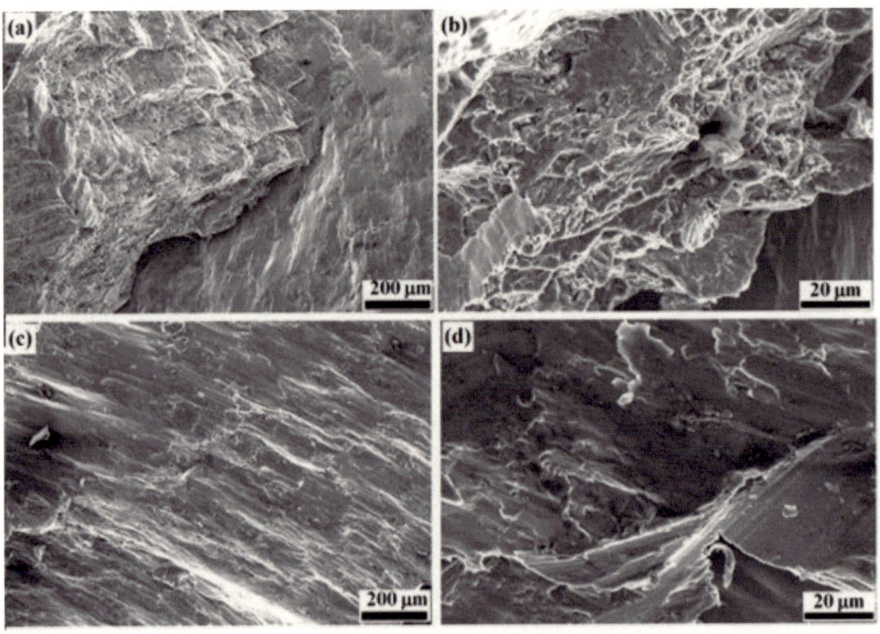

Figure 8.
SEM micrographs of the fracture surface of as-cast Mg matrix composites tested at room temperature (a) and (b): PL700 composite and (c) and (d): PL800 composite [1].

representing brittle fracture. It can be understood clearly that the mechanical properties (CYS, UCS and ductility) of as-cast composites can be improved significantly if one could avoid the formation of brittle Mg_2Si ceramic phase in the molten magnesium. This can be achieved by lowing the process temperature from 800 to 700°C during in-situ pyrolysis. Such improvement in mechanical properties of the final composites arises because of two primary reasons namely (i) minimal amount of grain boundary segregation of PDC particles tending to provide reasonable strengthening of the magnesium matrix, and (ii) reduction in the amount of Mg_2Si crystal within the magnesium matrix leads to cause significant plastic deformation of the composites [1].

4. Structure-property correlation of as-cast In-situ Mg-alloy based metal matrix composites

4.1 Microstructural characteristics

Figure 8 shows the microstructural evolution of as-cast PP900-AZ specimen. It can be evident that the intermetallics of β-$Mg_{12}Al_{17}$ phase are segregated at the grain boundaries as seen in **Figure 9(a)**. Most interestingly, majority of the SiCNO particles (black color) are entrapped within discontinuous network of β-$Mg_{17}Al_{12}$ phase at the vicinity of grain boundaries of PP900-AZ specimen as shown in **Figure 9(b)**. **Figure 9(d)** represents EDS spectra showing the presence of Si, C, and O atoms along with Mg and Al peaks. Microstructural analysis reveals no signature of Mg_2Si particles in the PP900-AZ specimen fabricated at 900°C. This is because of the fact that the existence of large cluster of Al-atoms (of about 9 wt%) in the molten slurry leads to maximize the probability of interrupting the diffusion path for Si-atoms to form any Mg_2Si crystals on heterogeneous substrates of SiCNO particles. This explanation is justifiable due to slower inter-diffusion rate of Al-atoms in the Mg solution as reported by Brennan et al. [1, 26]. However, Sachin et al. [27] observed the formation of in-situ Mg_2Si ceramic phase in the ultrasonic agitated molten AZ91 Mg-alloy after the addition of Si particles. It should be kept in mind that the polymer precursor approach does not involve any ultrasonic assisted vibration treatment of the molten Mg-alloy [1]. Yang et al. [28] mentioned that an ultrasonic vibration

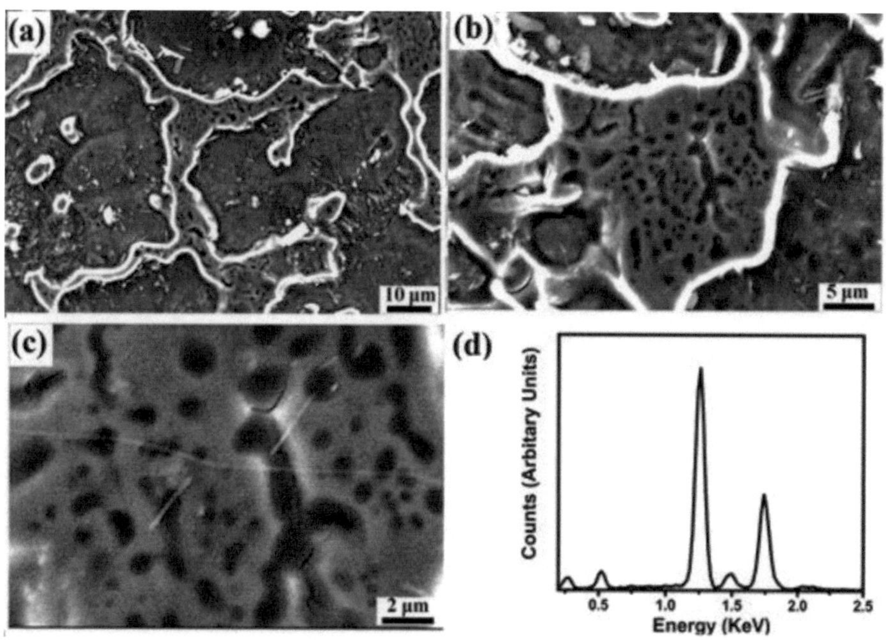

Figure 9.
Microstructural evolution of AZ91 matrix composites (a) β-$Mg_{12}Al_{17}$ intermetallics at the grain boundaries (b) and (c) encapsulation of SiCNO particles within the β-$Mg_{12}Al_{17}$ phase, and (d) EDS spectrum of polymer derived ceramic (SiCNO) particles [1].

can produce transient micro "hot spots" that can take temperature of about 5000°C and pressure above 1000 atmospheres in the melt [1]. Such a drastic variation in temperature–pressure accelerates the reaction kinetics of Mg_2Si formation as explained by Sachin et al. [27]. In addition, Sachin et al. [27] introduced the native powder of Si particles into the Mg-alloy melt which results in intimate physical contact between Si particles and the Mg melt. However, Si-atoms are introduced in the form of cross-linked polymer in the polymer precursor approach [1].

During solidification, molten Mg-alloy can be first transformed into primary α-Mg and β-$Mg_{17}Al_{12}$ phases in accordance with phase diagram. The primary α-Mg phase has limited amount of solubility with Al-atoms depending up on the temperature (maximum solubility of 11.8 at% Al-atoms at 437°C to 1 at% at room temperature). Therefore, Al-atoms have a greater tendency to push away any SiCNO particles to the grain boundaries which eventually leading to particle entrapment by β-$Mg_{12}Al_{17}$ phase. Hashim et al. [29] pointed out that grain boundary segregation of SiC particles occurs in Al-based MMCs owing to poor wettability between Al melt and SiC particles. Despite the fact that SiC and SiCNO particles are chemically distinct, it is justifiable to consider both of them as equivalent in terms of wettability properties with Al-atoms. The formation of Mg_2Si crystals was suppressed as most of the SiCNO particles are entrapped by β-$Mg_{17}Al_{12}$ phase. Under this situation, diffusion of Si-atoms from SiCNO ceramic phase could not take place across the domains of supersaturated α-Mg and β-$Mg_{17}Al_{12}$ phases during solidification. Therefore, the probability of forming Mg_2Si crystal within PP900-AZ composite can be ruled out completely [1].

Figure 10 represents the microstructural characteristics of as-cast PP900-AE specimen. As shown in **Figure 10**, it can be observed that Mg_2Si crystals exhibit dendritic morphology (average particle size of 50–100 μm) along with dispersion of Al_xRE_y intermetallics (acicular shaped gray color particles) in the matrix. The particle size of the fewer Al_xRE_y precipitates are appeared to much finer in size

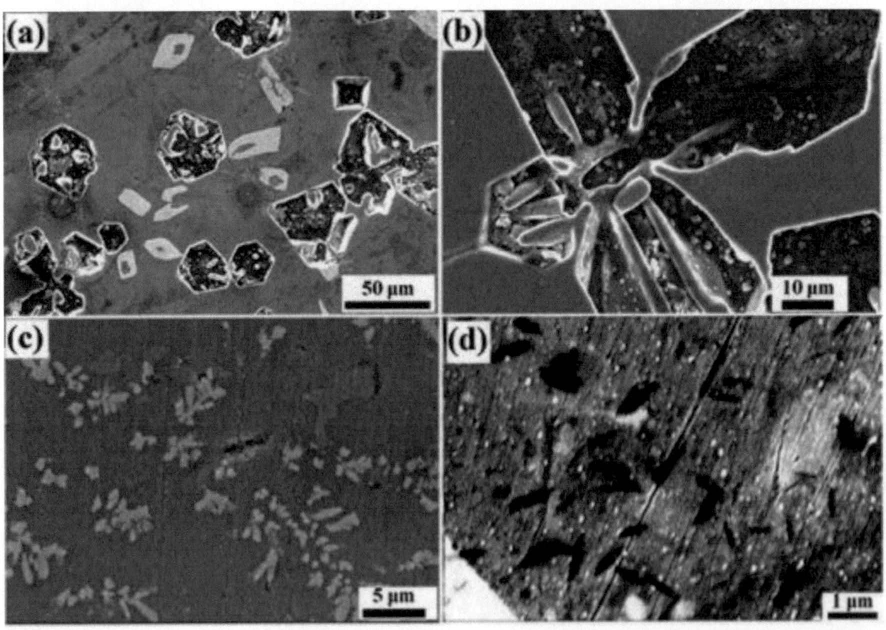

Figure 10.
Microstructural evolution of AE44 matrix composites (a) dispersion of coarsened Mg_2Si crystals and Al_xRE_y intermetallics (b) dendritic morphology of Mg_2Si crystal (c) dispersion of fine-sized Al_xRE_y intermetallics and (d) uniform dispersion of fine-sized SiCNO particles [1].

DOI: http://dx.doi.org/10.5772/intechopen.94305

as indicated in **Figure 10(c)**. It can be seen that SiCNO particles (a width of 100 to 200 nm and a length of 0.5 to 1 μm) are distributed homogenously throughout the matrix (**Figure 10(d)**). During in-situ pyrolysis, the chance of forming Mg_2Si crystals seems to be limited again for the same reasons mentioned earlier for PP900-AZ composite [1]. However, the surrounding medium for nucleating Mg_2Si crystals is completely different during solidification [1]. In AE Series Mg-alloy, the liquid phase was converted in to primary α-Mg phase and Al_xRE_y phase. Most of the Al-atoms are expected to chemically bond with RE elements in the molten Mg-alloys to form an acicular Al_xRE_y precipitates. Moreover, Chen et al. [30] found that addition of RE elements have positive effects on the nucleation of Mg_2Si crystals for the case of Al-Mg-Si alloys. Therefore, the most preferential sites for nucleating Mg_2Si crystals could be the Al_xRE_y precipitates in AE-44 Mg-alloy [1]. Once Mg_2Si crystal is nucleated, and it persists its growth along the preferential growth direction **<100>** to form an equilibrium crystal shape of octahedron morphology. However, the morphology of Mg_2Si crystal changes from octahedral to dendritic shaped crystal because of SiCNO particles in the surrounding medium (**Figure 10(b)**) which may impose space constraint for this equilibrium growth direction (**Figure 10(c)**).

4.2 Mechanical properties

As seen in **Table 2**, the micro-hardness and yield strength data of the fabricated composites reveals that PP900-AE composite experiences hardness or strength enhancement with an increment of 332.0% as compared to that of the base AE44 Mg-alloy. However, there is only marginal increment in the hardness or yield strength of PP900-AZ composite as compared to the unreinforced AZ91 Mg-alloy. The possibility of impeding the movement of dislocation within the primary α-Mg phase seems to be futile as most of the SiCNO particles are entrapped within discontinuous network of β-$Mg_{17}Al_{12}$ phase (**Figure 8(b)**) [1]. Nevertheless, PP900-AZ composite is expected to offer high temperature creep resistance as unstable β-$Mg_{17}Al_{12}$ phase is dispersed with thermally stable SiCNO particles. The observed significant improvement in hardness for PP900-AE composites is mainly associated with the contribution of Taylor strengthening and load transfer mechanisms. The contribution of Taylor strengthening is massive in case of PP900-AE specimen as compared to that of PP900-AZ specimen. Notice that all of the fabricated in-situ P-MMCs were subjected to temperature difference (ΔT) of about 873 K during solidification processing. Therefore, the probability of achieving highest density of dislocations increases steadily with fabricated composites owing to larger thermal strain induced during solidification. However, thermal mismatch effect loses its importance in first place for the case of PP900-AZ specimen. This can be associated with the entrapment of SiCNO particles by β-Mg17Al12 phase, which in turn generates lesser number of dislocations within the primary α-Mg

	AZ91	AE44	PP900-AZ	PP900-AE
Vicker's hardness, HV	82	58	95	254
Yield strength, MPa	288	206	310	880
Volume fraction of SiCNO	0.0		0.05	0.10
Volume fraction of Mg_2Si	0.0		0.0	0.18

Table 2.
Mechanical properties of the fabricated In-situ Mg-alloy based composites.

phase which eventually leads to a lowest Taylor strengthening. Next to Taylor strengthening, the load transfer strengthening also plays a major role in enhancing the hardness of the fabricated composites. This can be mainly associated with load bearing capacity of hard ceramic phase due to presence of higher volume fraction of reinforced PDC particles (i.e. $V_f \approx 0.28$) in PP900-AE specimen [1].

5. Conclusions

Commercial purity magnesium and Mg-alloys (AZ91 & AE44 series) based composites were fabricated successfully by injecting liquid polymer and cross-linked polymer directly into molten metal/alloys by using liquid stir-casting method via polymer precursor approach. During in-situ pyrolysis, the intensity for formation of brittle Mg_2Si ceramic phase can be minimized by reducing the processing temperature from 800 to 700°C. Most of the polymer derived SiCNO particles are segregated at the grain boundaries because of particle pushing effect by solidification front during solidification. In-situ Mg_2Si crystals were not formed in AZ91 matrix composite because of higher concentration and slower inter-diffusion rate of Al-atoms within molten slurry. To summarize, it may be emphasized that in-situ pyrolysis approach opens an opportunity window for the material researchers to develop futuristic Mg-based hybrid in-situ MMCs owing to the formation of both the in-situ SiCNO and Mg_2Si phases. Furthermore, in-situ pyrolysis produces a wide range of particle size, and morphologies of PDC particles, so that one can tailor the mechanical properties for specific engineering applications. Notice that ductility of the composites can be enhanced by controlling the volume fraction, size and morphologies of Mg_2Si crystals in the final composites. This can be achieved via optimization of in-situ processing variables such as pyrolysis or process temperature, pyrolysis time, and cooling rate of casting process.

Acknowledgements

The authors would like to express their gratitude to Prof. Rishi Raj, University of Colorado, Boulder, for his valuable inputs, and for providing the precursor materials under Grant No. DMR1105347 supported by the National Science Foundation. The authors also would like to thank Indian Institute of Technology Ropar and Indian Institute of Science Bangalore for providing permission to complete this research work.

Author details

Nagaraj Chelliah Machavallavan[1]*, Rishi Raj[2] and M.K. Surappa[3]

1 PSG College of Technology, Coimbatore, 641004, India

2 University of Colorado, Boulder, 80309, USA

3 Anna University, Chennai, 600025, India

*Address all correspondence to: cmnmett@gamil.com

References

[1] Nagaraj C.M, Processing and characterization of in-situ magnesium metal matrix composites containing SiCNO particles, [Ph.D Thesis]. Ropar : Indian Institute of Technology; 2017.

[2] Koczak M.J, and Kumar K.S, In-situ process for producing a composite containing refractory material. US Patent 4,808,372, (1989).

[3] Koczak M.J, and. Premkumar M.K, Emerging technologies for the in-situ production of MMCs. JOM-J. Min. Met. Mater. Soc., 1993 : 45 :44-48.

[4] Gibbons T.B, and O'hara S, The effect of internal oxidation on the damping capacity of copper-silicon alloys. J. Phil. Magaz., 1960 :5 : 140-145.

[5] Tjong S.C., and Ma Z.Y, Microstructural and mechanical characteristics of in-situ metal matrix composites, Mater. Sci. Eng. R., 2000 : 29 : 49-113.

[6] Daniel B.S.S, Murthy V.S.R, and Murty G.S, Metal-ceramic composites via in-situ methods, J. Mater. Process. Tech.,1997:68:132-155.

[7] Pradeep K Rohatgi, Ajay Kumar P, Nagaraj M Chelliah, Rajan T.P.D, Solidification processing of cast metal matrix composites over the last 50 years and opportunities for the future, Journal of Metals, 2020 :1:1-15.

[8] Chelliah N.M, Padaikathan P., Surappa M.K., Effects of processing conditions on solidification characteristics and mechanical properties of in situ magnesium metal matrix composites derived from polysilazane precursor, Journal of Composite Materials, 2019 :1 :1-15.

[9] Chelliah N.M, Padaikathan P, Surappa M.K, Deformation mechanisms and texture evolution of in- situ magnesium matrix composites containing polymer derived SiCNO dispersoids during hot compression, Material Science and Engineering: A, 2018 :720:49-59.

[10] Chelliah N.M, Sudarshan, Lisa Kraemer, Singh H, Surappa M.K, Rishi Raj, A polymer route to the design of thermally stable metal matrix composites: Materials selection and in-situ processing, Research and Report on Metals, 2017 :1 :3-7.

[11] Chelliah N.M, Sudarshan, Lisa Kraemer, Singh H., Surappa M.K, Rishi Raj, Stress-rupture measurements of cast magnesium strengthened by in-situ production of ceramic particles, Journal of Magnesium and Alloys, 2017 :5 : 225-230.

[12] Chelliah N.M, Singh H, Surappa M.K., Microstructural evolution and strengthening behavior in in- situ magnesium matrix composites fabricated by solidification processing, Material Chemistry and Physics, 2017 : 194 :65-76.

[13] Chelliah N.M, Singh H, Raj R, and Surappa M.K, Processing, microstructural evolution and strength properties of in-situ magnesium matrix composites containing nano-sized polymer derived SiCNO particles, Material Science and Engineering A, 2017 :685 :429-438 DOI:10.1016/j. msea.2017.01.001 (I.F: 3.086)

[14] Raj R, Surappa MK, Process for preparation of nano ceramic-metal matrix composites and apparatus thereof US Patent 8,540,797

[15] Sudarshan, Surappa M. K, Ahn D, and Raj R, Nanoceramic-metal matrix composites by in-situ pyrolysis of organic precursors in a liquid melt, Met. Mater. Trans A., 2008 :39:3291-3297.

[16] Sudarshan, Terauds K, Anil Chandra A. R., and Raj R, Polymer-derived in-situ metal matrix composites created by direct injection of liquid polymer into molten magnesium, Met. Mater. Trans. A., 2014) :45 :551-554.

[17] Castellan E, Ischia G, Molinari A, and Raj R, (2013) A novel in-situ method for producing a dispersion of a ceramic phase into copper that remains stable at 0.9 Tm. Mett. Mater. Trans. A., 2013 : 44 :4734-4742.

[18] Castellan E, Kailas S.V, Madayi S, and Raj R, Low-wear high-friction behavior of copper matrix composites dispersed with an in-Situ polymer derived ceramic, J. Tribology.,2015:137:024501.

[19] Kumar A, Raj R, and Kailas S.V, A novel in-situ polymer derived nano ceramic MMC by friction stir processing, Mater. Des., 2015: 85:626-634.

[20] Kroke E, Li Y.L, Konetschny C, Lecomte E, Fasel C, and Riedel R, Silazane derived ceramics and related materials. Mater. Sci. Eng. R, 2000 :26 : 97-199.

[21] Yuan X, Sun W, Du Y, Zhao D.D and H. Yang, Thermodynamic modeling of the Mg–Si system with the Kaptay equation for the excess Gibbs energy of the liquid phase, Calphad., 2009 :33 : 673-678.

[22] Inem B, and Pollard G., Interface structure and fractography of a magnesium-alloy, metal-matrix composite reinforced with SiC particles, J. Mater Science,1993 :28 :4427-4434.

[23] Kurtz W, and Fisher D.J, Fundamentals of solidification, 1st ed, Switzerland, Trans Tech Publication ; 1986.

[24] Kim G.H, Hong S.M, Lee M.K, and Kim I.S, Effect of oxide dispersion on dendritic grain growth characteristics of cast aluminum alloy, Mater. Trans., 2010 : 51:1951-1957.

[25] Sudarshan, Magnesium matrix-nano ceramic composites by in-situ pyrolysis of organic precursors in a liquid melt [Ph.D Thesis]. Bangalore : Indian Institute of Science ; 2010.

[26] Brennan S, Warren A.P, Coffey K.R,. Kulkarni N, Todd P,. Kilmov M, and Sohn Y.H, Aluminum Impurity Diffusion in Magnesium, J. Phase Equil. Diff., 2012 :X :121-125.

[27] Sachin V.M, Singh S.P, Sinha P, Bhingole P, and Chaudhari G.P, Microstructural evolution in ultrasonically processed in situ AZ91 matrix composites and their mechanical and wear behavior, Mater. & Des., 2014 : 53:475-481.

[28] Yang Y, Li X, and Cheng X, Ultrasonic assisted fabrication of metal matrix nanocomposites, J. Mater. Sci., 2004 :39 :3211-3212.

[29] Hashim J, Looney L, and Hashmi M.S.J, The wettability of SiC particles by molten aluminum alloys, J. Mater. Proc. Tech., 2001 :11 :9324-328.

[30] Chen Z, Hao X, Zhao J. and Ma C, Kinetic nucleation of primary a(Al) dendrites in Al-7%Si-Mg with Ce and Sr additions, Trans. Non. Ferr. Met. Soc. China., 2013 :23 :3561-3567.

Section 3

Modern Methods of Magnesium Alloys and Magnesium-Based Composites Structure Forming

Chapter 7

Synthesis of Nano-Composites Mg_2TiO_4 Powders via Mechanical Alloying Method and Characterization

Ranjan Kumar Bhuyan, Bhagban Kisan, Santosh Kumar Parida, Soumya Patra and Sunil Kumar

Abstract

In this chapter a systematic investigation of impact of mechanical activation on structural, microstructural, thermal and optical properties of $MgO - TiO_2$ nanocrystalline composite system, synthesized via high energy ball milling techniques. Williamson-Hall (W-H) plot method was employed to understand the signature of the broadening in the XRD peaks and for the estimation of crystallite size of $MgO - TiO_2$ nanocrystalline composite system. It revealed that the peak broadening is not only due to reduced coherently diffracting domain size but also due to a significant strain distribution. The calculated strain was 9.01×10^{-3} and the average crystallite sizes were 40–60 nm for 35 hours (hrs) milled powder and this result is very much consistent with transmission electron microscopy (TEM) analysis. The SAED ring pattern indicates that the phase of Mg_2TiO_4 - nanoparticles was polycrystalline in structure and the distance between crystalline planes was consistent with the standard pattern for a spinel Mg_2TiO_4 crystal structure. To analyze the lattice fringes for the 35 hrs milled samples high resolution-TEM (HR-TEM) study was carried out and the result revealed that the each particle has single crystalline structure. Morphological studies were carried out by using SEM analysis. The thermal decomposition behavior of the milled powders was examined by a thermogravimetric analyzer (TGA) in argon atmosphere. Also, MTO nanoparticles showed a strong absorption at ~356 nm and the band gap values ranged between 3.26-3.78 eV with an increase of milling time from 0 to 35 hr. The mechanically derived MTO nanoparticles showed promising optical properties which are suitable for commercial optoelectronic applications.

Keywords: high energy ball milling, $MgO - TiO_2$ composite system, W-H method, microstructure, TEM and HR- TEM, DSC-TGA analysis, optical studies

1. Introduction

The tremendous growth in telecommunication industries has led increasing demand on the development of low loss and low cost high frequency dielectric ceramics in the form of resonators, filters, antennas, substrates [1–4]. Moreover, the recent demand has focused on searching for low loss materials with lower permittivity to miniaturize the devices [4–7]. This is because low permittivity can not only reduce cross coupling with conductors but also precise the time for the transmission of electronic signal. Moreover, due to extension of carrier frequency high-quality factor ($Q \times f_o$) and nearly zero temperature coefficient of resonant frequency (τ_f) also play crucial roles for frequency selectivity and temperature stability of the system, respectively [5–10]. Various smart electronic materials were proposed to fulfill these requirements for high-frequency applications [5–13]. Recently, searching for low loss dielectric ceramics based on MgO - TiO_2 binary system has brought much more attention. It was reported that MgO – TiO_2 system has three stable phases, such as $MgTiO_3$, Mg_2TiO_4 and $MgTi_2O_5$ [14, 15], which are used for microwave engineering/RF- applications. These binary magnesium titanates ($MgTiO_3$, Mg_2TiO_4 and $MgTi_2O_5$) are differed extremely from other materials due to their good dielectric properties. It has been established that $MgTiO_3$ possesses ilmenite structure, $MgTi_2O_5$ has pseudobrookite structure and Mg_2TiO_4 has an inverse spinel structure belonging to cubic space group of Fd3m (227) [16, 17]. Out of them magnesium orthotitanate (Mg_2TiO_4) is a promising dielectric material with excellent microwave dielectric properties, i.e., medium dielectric constant (ε_r) ~ 14, high-quality factor ($Q \times f_o$) ~150,000 GHz and a negative temperature coefficient of resonant frequency (τ_f) ~ – 50 ppm/°C [15].

Ceramic nonmaterials have great scientific interest due to their unique physical and chemical properties and are significantly different from bulk counterparts. A bulk material has fixed physical properties regardless of its size, but reducing the particle size into nanoscale, by keeping chemical composition fixed, can change the fundamental properties of the materials [18]. A unique aspect of nanoscale materials is that they have large surface area to volume ratio, which opens new possibilities of surface dependent phenomena that are practically very useful for various applications. As the size of the material decreases into the nanoscale dimensions (less than 100-200 nm), a number of physical phenomena have come into notice, which drove our attention for the synthesis of nanocrystalline - Mg_2TiO_4 powders. There are very few papers are available about the effect of mechanical activation on MgO - TiO_2 binary system and to investigate its physical changes. Recently, Bhuyan et al., [19], have studied the influence of high energy ball milling on structural, microstructural and optical properties of Mg_2TiO_4 nanoparticles. They proposed that MTO nanoparticles prepared by mechanical alloying method exhibited promising optical properties which are suitable for commercial optoelectronic applications. In another study, Bhuyan et al., [20], have studied the structural and microwave dielectric properties of Mg_2TiO_4 ceramics synthesized by mechanical alloying method. Cheng et al. [21], have investigated the microwave dielectric properties of Mg_2TiO_4 ceramics synthesized via high energy ball milling method. Filipovic et al., [22], have studied the influence of mechanical activation on microstructure and crystal structure of sintered MgO-TiO_2 system.

In this present chapter, Mg_2TiO_4 (MTO) nano-composite ceramics were synthesized via mechanical alloying (MA) method with the help of high energy planetary ball milling. Mechanical alloying is a most efficient, cost effective and convenient method for the synthesis of a wide range of nanosized metallic and ceramic powders [23]. This method has many advantages such as simplicity, relatively inexpensive compared to other techniques to produce large scale nanoparticles and can be applicable to any type of materials [24]. The most important merit of this technique is that the solid-state reaction is activated via mechanical energy rather than production of heating energy. Moreover, mechanically synthesized powders have good physical properties than those derived by a conventional solid-state reaction and most of the wet-chemical processes [25]. Mechanical synthesis not only makes the material finer but also includes structural changes, phase transformations and even solid-state reactions among the solid reagents. These physicochemical changes occur due to the efficient transformation of the mechanical energy of the grinding media to the particles and the intensive mechanical force during the milling process [26].

It is well known that a perfect crystal would extend in all possible directions to infinity; however, no such crystals are perfect due to their finite size. This deviation from its perfect crystallinity is the main cause for broadening of the X-ray diffraction peaks of materials. There are two important characteristics extracted from the peak width analysis viz. crystallite size and lattice strain. Crystallite size is a measure of the size of a coherently diffracting domain whereas lattice strain is a measure of the distribution of lattice constants arising from crystal imperfections, such as lattice dislocation. The other sources of strain are the grain boundary triple junctions, crystal imperfections, contact or sinter stresses, stacking faults etc. [26]. The X-ray line broadening is used for the investigation of dislocation distribution. Moreover, it should be noted that crystallite size of the particles is not same as the particle size due to the presence of polycrystalline aggregates. The particle size can be measured from various techniques such as scanning electron microscope (SEM) or field emission scanning electron microscope (FE-SEM) and transmission electron microscope (TEM) analysis. Various methods are adopted by different researchers for the estimation of crystallite size and lattice strain, which are X-ray peak profile analysis (XPPA), pseudo-Voigt function, Rietveld refinement, and Warren-Averbach analysis [27–29]. However, in the present study, Williamson–Hall (W–H) method is a simplified integral breadth method employed for the determination of crystallite size and lattice strain, by considering the peak width as a function of 2θ [30].

In this chapter, the impact of mechanical activation of the MgO-TiO_2 system for the synthesis of nanocrystalline Mg_2TiO_4 powders via high energy ball milling technique has been investigated. The effect of milling time on crystal structure, microstructure, thermal and optical properties of this proposed system is being studied. This study further reveals the importance of W- H method for the determination of crystallite size and lattice strains.

2. Experimental details

2.1 Materials

The Mg_2TiO_4 powders were prepared from commercially available high- purity oxides MgO (99.99% purity) and TiO_2 (99.99% purity) of Sigma Aldrich (St.Louis, MO), as starting materials.

2.2 Synthesis of Mg_2TiO_4 nano-composite alloys

In this chapter, high energy planetary ball milling techniques was used for the synthesis of Mg_2TiO_4 nano-composite alloys from high- purity MgO and TiO_2 oxides. The starting materials were weighted according to desired stoichiometry ratios and milled for 5 – 35 hrs to reach steady state condition using planetary ball mill (Fritsch, GmBH, Germany) with the following parameters: (i) ball-to-powder ratio: 10:1; ball diameter: 8 and 16 mm; ball and vial materials: harden stainless steel; the vial rotation speed: 350 rpm. In order to avoid significant temperature rise, the milling process was stopped periodically for every 10 minutes and then resumed for 5 minutes. A brief description about the ball milling techniques is summarized below.

Ball milling technique: The photographic view of planetary ball mill is depicted by **Figure 1(a)** and in **Figure 1(b)** it is shown the Zirconia jars with zirconia balls. The zirconia balls are selected in order to avoid the contamination with the given sample. The vials mounted on the horizontal disc rotate in a direction opposite to that of the disc and thereby simulating a planetary-like motion (as shown in **Figure 2**). This result in large outward force acting on the balls kept inside the vial and causes the balls to collide with them and also to the wall of the vial more energetically. When the Mg_2TiO_4 oxide powders are kept inside the vial along with the balls, the powders undergoes repeated cold welding and fracture at the surfaces of the balls and the vial. This process leads to disintegration of the powders, resulting first in refinement of crystallite size to produce nanocrystalline alloys along with a large number of defects in the parent powders [26, 31–33]. Hence, crystallite size refinement is a natural consequence of a ball milling process. As the milling time progresses the alloy becomes amorphous. The refinement and alloying processes are estimated by the milling parameters including ball to powder weight ratio, size of the ball, speed of rotation, duration of milling etc. Moreover, the nature of the milling container (or vial), milling media and types of balls used during the milling process also played an important role in synthesizing nanocrystalline powders [34].

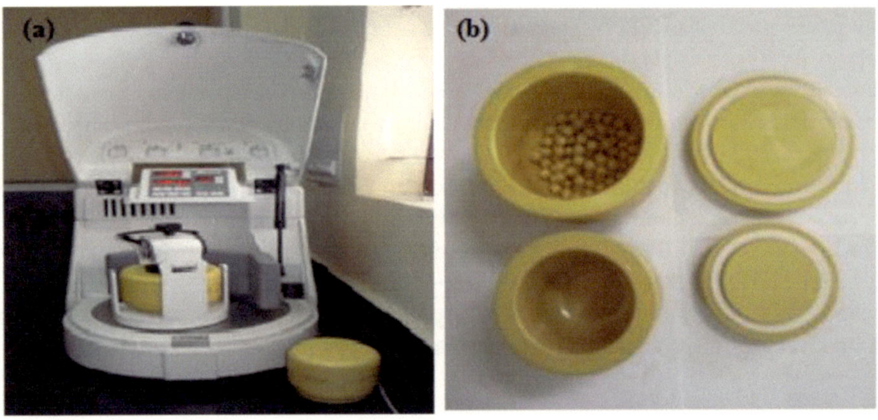

Figure 1.
(a) Photographic view of planetary ball mill (b) zirconia jar with zirconia balls.

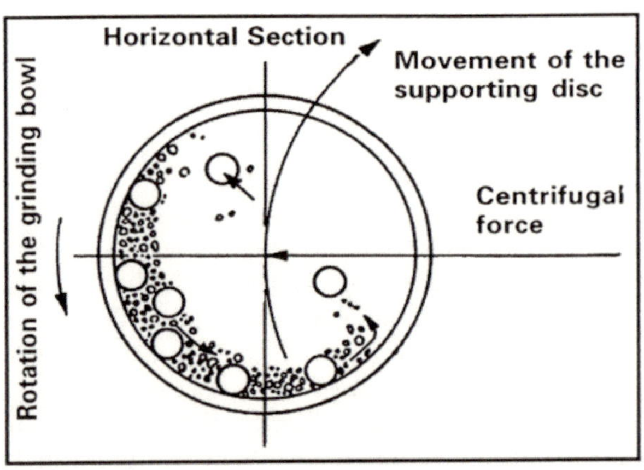

Figure 2.
Schematic diagram of the horizontal section of a vial depicting the movement of the balls inside the vial due to its planet like movement [26].

2.3 Characterizations techniques

Room temperature X-Ray diffraction (XRD) analysis was carried out to determine the crystallite size, lattice strain, lattice parameter and phase analysis of the different hours milled powder sample using X-ray Diffractometer (Rigaku TTRAX 18 KW CuK_α radiation). Recording of microstructure of 35 hrs nanocrystalline Mg_2TiO_4 milled powders has been carried out by using transmission electron microscope (TEM, JEOL 2100; JEOL, Tokyo, Japan). The surface morphology of nanocrystalline Mg_2TiO_4 milled powders was observed by scanning electron microscopy (SEM, Leo 1430, PV, Carl Zeiss Jena, Germany). The thermal decomposition behavior of the different hours milled powders was examined by a thermogravimetric analyzer (TGA, NETZSCH, STA 449-F3. Jupiter) at a heating rate of 10°C/min in the argon atmosphere. UV–VIS–NIR spectrophotometer (UV 3101PC, Shimdzu) was used to obtained UV/VIS absorption spectra of all the samples in the wavelength range 200–1000 nm.

3. Results and discussion

3.1 Structural properties

X-ray diffraction (XRD) patterns of prepared Mg_2TiO_4 nanoparticles milled for different hours via high energy ball milling was taken and is illustrated in **Figure 3**. XRD patterns indicates that for the samples milled for 5 hrs exhibited the peaks corresponding to initial compounds MgO and TiO_2 only. When the milling time increases, the intensities of the parent oxide peaks appear to be depressed gradually and the formation of associated $MgTi_2O_5$ phase was observed. When the milling time increased to 30 hrs, all the starting oxides peaks are disappeared completely. At the same time high intense diffraction peaks of pure- Mg_2TiO_4 phase are observed with small significance of $MgTiO_3$ and $MgTi_2O_5$ phases. However, when the milling time reached to 35 hrs, the

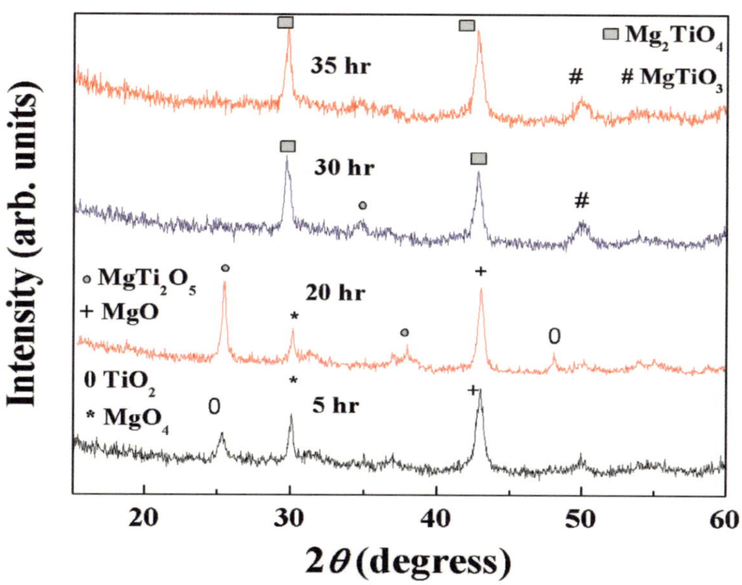

Figure 3.
XRD patterns of the MgO and TiO_2 oxides milled for 5, 20, 30 and 35 hrs (adapted with permission from Bhuyan et al., 2020, @ Springer [19]).

sample showed more intensified Mg_2TiO_4 phases along with small $MgTiO_3$ peak. This signifies that crystallite nature of pure Mg_2TiO_4 sample enhances and is confirmed from ICSD – PDF # 06–5792. The presence of associated phases such as $MgTi_2O_5$ and $MgTiO_3$ in the MgO - TiO_2 system is mostly due to the difference in the degree of the incipient mechanical reaction. This can be explained as follows: at the time of milling, the mechanical energy of the grinding media transforms into the given oxide particles that causes structural destruction followed by reduction in particle size [26].

3.2 Williamson - Hall (W - H) method

The crystallite size of nanoparticles can be determined with several techniques that rely upon the peak width of the X - ray diffraction patterns. In the present study, Williamson-Hall (W-H) plot method as well as Scherrer formula have been chosen in order to understand the origin of the broadening in the XRD peak.

The broadening of XRD peaks is due to crystallite size and strain contributions. The average crystallite size was calculated from XRD peak width based on Debye–Scherrer's equation,

$$D = \frac{k\lambda}{\beta_{\mathrm{hkl}} \cos\theta} \tag{1}$$

where β_{hkl} is the full width half maximum, D is the crystallite size, k is the shape factor which is taken as 0.9 for spherical particles, λ is the wavelength of incident X-ray radiation (λ = 0.154 nm for Cu-K_α) and θ is the Bragg angle of the analyzed peaks.

According to Williamson and Hall, the strain-induced broadening in nanocrystalline powders due to crystal imperfection and distortion was calculated using the formula [35],

$$\varepsilon = \frac{\beta_{\mathrm{hkl}}}{4\tan\theta} \tag{2}$$

Here, ε is the effective strain associated with mechanical alloying. Now, the total peak broadening is defined as the sum of the contributions of crystallite size and strain present in the material and can be expressed as [36],

$$\beta_{\mathrm{hkl}} = \beta_D + \beta_\varepsilon \tag{3}$$

where β_D is due to the contribution of crystallite size, β_ε is due to strain-induced broadening and β_{hkl} is the width of the half-maximum intensity of instrumental corrected broadening. This β_{hkl} can be calculated by using the relation,

$$\beta_{hkl} = \left[\left(\beta^2{}_{hkl}\right) Measured - \left(\beta^2{}_{hkl}\right) Instrumental\right]^{1/2} \tag{4}$$

If we consider the particle size and strain contributions to line broadening are independent to each other and both have a Cauchy-like profile, then the observed line breadth is the sum of Eqs. (1) and (2) and is given by [35],

$$\beta_{\mathrm{hkl}} = \left[\frac{k\lambda}{D\cos\theta}\right] + 4\varepsilon\sin\theta \tag{5}$$

By rearranging the above equation, we get,

$$\beta_{\mathrm{hkl}}\cos\theta = \left[\frac{k\lambda}{D}\right] + 4\varepsilon\sin\theta \tag{6}$$

This is the Williamson- Hall equation, which represents the uniform deformation model. The average crystallite size is estimated for selected peaks of nanocrystalline MTO powders milled for different hours by using Eqs. (1) and (6). The variation of average crystallite size as a function of milling time calculated by both the method is plotted and is depicted in **Figure 4(a)**. It is clear from **Figure 4(a)** that up to 30 hrs of milling the average crystallite size decreases sharply and then attains a constant value. The average crystallite size of the parent sample was found to be nearly 2.5 μm. But for 20 hrs of milling, the crystallite size reduced to 100-120 nm and for 35 hrs of milling it becomes 40-60 nm, as calculated by W-H method. From Scherer formula the average crystallite size for MTO powder are found to be 28 nm and 17 nm, respectively after 20 and 35 hrs of milling. Thus the crystallite size calculated from the Scherer equation is smaller than that of the W-H method. This is due to the fact that the Scherer's equation does not account for the lattice strain effect that causes line broadening.

Mg_2TiO_4 has an inverse spinel structure and having structural formula $Mg[\text{MgTi}]O_4$ and belonging to the cubic space group of Fd3m (227) [15]. The Ti and Mg atoms occupy the tetrahedral (8a) and octahedral (16d) sites and the oxygen atoms are in (32e) site symmetry position [20]. According to Bragg's law [37],

$$2d\sin\theta = n\lambda \tag{7}$$

where, n is the order of diffraction and it is usually taken as $n = 1$, λ is the wavelength of incident X-ray and d is the spacing between parallel planes of given miller indices h, k *and* l. Since, Mg_2TiO_4 has cubic structure, so the lattice constants are $a = b = c$. The d-spacing is related to the lattice constant a, and the miller indices h, k *and* l, by the following relation [37],

$$d = \frac{a}{\sqrt{h^2 + k^2 + l^2}} \tag{8}$$

By using Eq. (7), the lattice constant of selected planes is calculated by following relation,

$$a = \frac{\lambda}{2\sin\theta}\sqrt{h^2 + k^2 + l^2} \tag{9}$$

The variation of lattice parameter as a function of different milling time is plotted and is illustrated as inset **Figure 4(a)**. The results showed that the lattice constant decreases with increase of milling time from 8.436 Å to a stable value of 8.412 Å. This difference in lattice constant stipulates the occurrence of atomic disorder due to the milling process. That means the grinding of the powders via high energy ball milling techniques not only reduces the crystallite size into nanoscale range (< 100 nm) but also causes in the enhancement of lattice strain. Thus, the net X-ray line broadening is due to decrease of crystallite size, development of lattice strains and also due to the instrumental effects. Normally, crystallite size is a measure of the size of a coherently diffracting domain. So, when the crystallites of the materials are <100 nm, they have very less number of parallel diffraction planes that causes broadened diffraction peaks. Similarly, the non-uniform strains arises out of heavy plastic deformation during the course of high energy mechanical milling process that causes broadening of the diffraction peaks [35].

The milling dependence of internal microstrain (e) of mechanically derived nanocrystalline -MTO powders was evaluated. The graph between $4\sin\theta/\lambda$ (taken along x-axis) and $\beta_{hkl}\cos\theta/\lambda$ (taken along y-axis) for selected diffraction

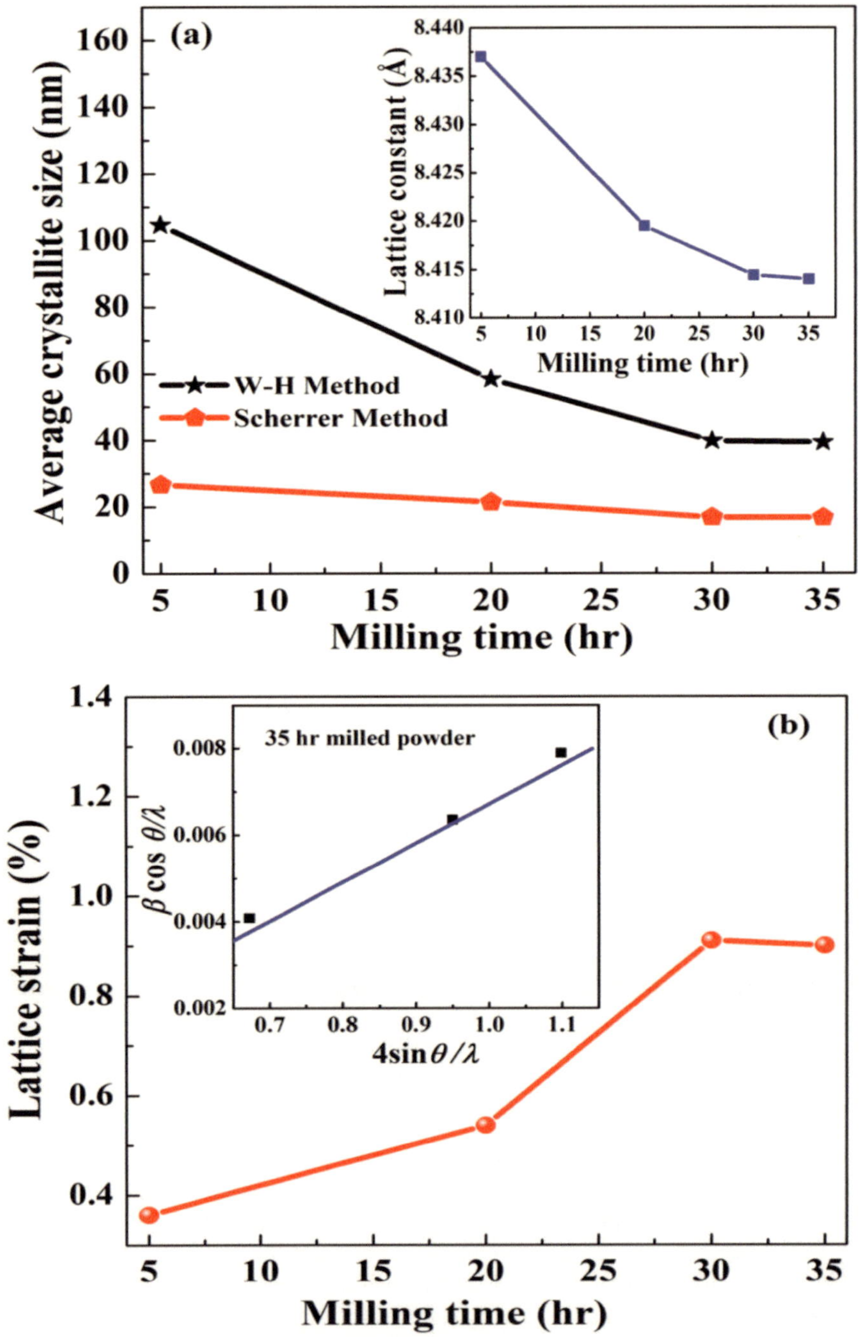

Figure 4.
*(a) Variations of average crystallite size and lattice parameter (**inset**) with milling time and (**b**) variation of lattice strain with milling time. **Inset:** W-H plot for 35 hrs milled powders. (adapted with permission from Bhuyan et al., 2020, @ Springer [19]).*

peaks for 35 hrs milled MTO nano-powders, is plotted and is depicted in inset **Figure 4(b)**. In the present case, the crystal is considered as isotropic in nature and it is assumed that the properties of material do not depend on the direction along which it is measured. From inset **Figure 4(b)**, (called W-H plot), it shows that the data points are not much deviated from the straight line suggesting isotropic nature of the strain. From the linear fit of the data, the average crystallite, *D* was estimated from the y-intercept and the microstrain (ε) from the slope of the fit (inset **Figure 4(b)**). For 35 hrs milled powders, the crystallite size is found to be approximately 38 nm and microstrain is around 9.01×10^{-3} respectively. Thus from the W-H analysis it is clear that the broadening of the X-ray peaks is due to the contribution of smaller crystallite size and the induction of strain. Further, it was noticed that with increase in milling duration the internal microstrain increases and it attains a constant value after a particular milling period.

3.3 Thermal analysis

In order to determine the characteristics temperature at which solid state processes are taking place, DTA and DSC analysis has been performed. **Figure 5** shows the DTA and DSC curves of the MTO system milled for 35 hrs. The initial weight loss of 6-8% was assigned to the evaporation of humidity during powder's preparation route. Around 400 -450°C, the weight loss was more prominent and apportioned to the formation of secondary phases of $MgTiO_3$ and $MgTi_2O_5$. As mechanical activation supports hygroscopy, so weight loss is the highest for the sample with the longest period of milling. So, MTO-35 has more weight loss compared with other different hours milled samples. Also, we have observed some endothermic and exothermic peaks that are related to the weight loss of the sample with temperature.

3.4 Morphological study on mechanically alloyed nanocrystalline MTO powders

3.4.1 SEM analysis

SEM micrographs were obtained to see the influence of mechanical activation on the evolution of microstructure of MTO powders milled for different hours (**Figure 6**). It is well-known that milling processes yield a significant modification

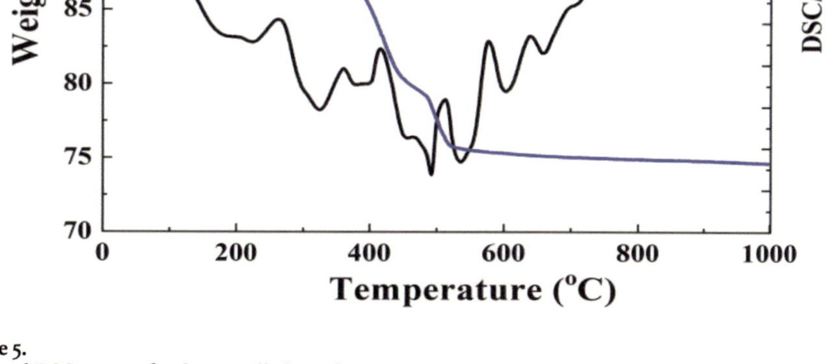

Figure 5.
TGA and DSC curves of 35 hours milled powders.

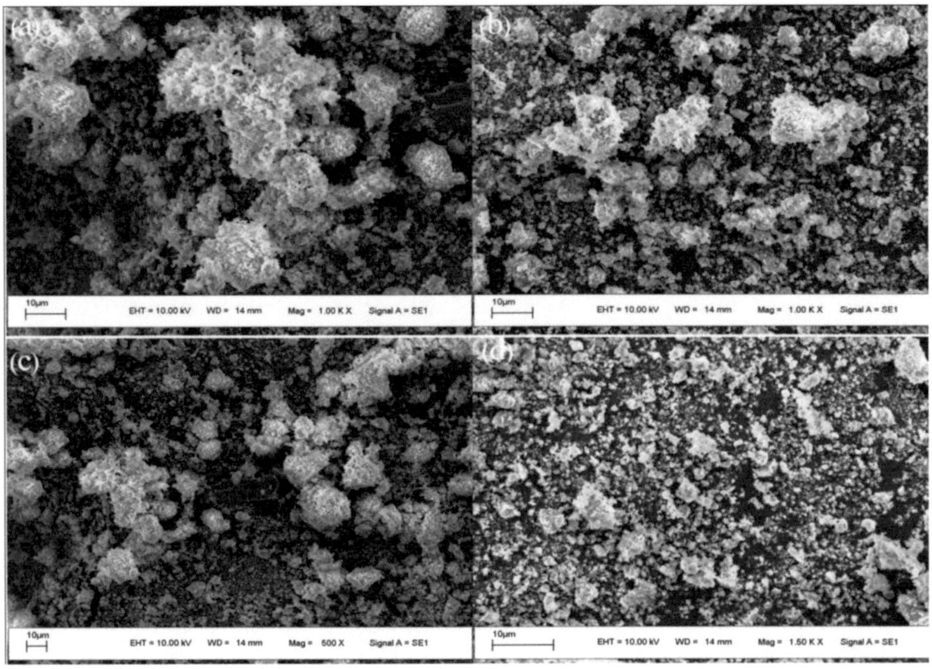

Figure 6.
SEM micrographs of the MTO powders milled for (a) 0 (b) 5 (c) 20 and (d) 35 hrs (adapted with permission from Bhuyan et al., 2020, @ Springer [19]).

in the morphology of composite materials due to severe plastic deformation of the particles during the milling process [38, 39]. Generally, the microstructure evolution is controlled by the processing parameters, such as composition of the materials, rise of temperature and milling intensity. From the morphological study, it is noticed that the starting powders consists of spherical particles with extreme agglomerated morphology. After a short period of milling up to 5 hrs, there is significant effect on the morphology of the MTO powders was observed. The particles were distributed over a wide range from sub-micrometer to few micrometers with spherical in morphology. As the milling time increased up to 35 hrs, cold welding of particles was activated and the size of the particles reduced into nanometer range due to the high impact collision of the balls. At this stage more distinct granular structure particles are observed as compared to the initial stages of milling along with the presence of new phases in the shape of agglomerates covered with many smaller nanosized particles of starting powders. These clustering of MTO nanoparticles are typically mechanically alloyed powders that are resulted from repeated cold welding and fracture of powders during the process of high energy mechanical alloying. The surface morphology of the nanocrystalline MTO powders are in support as evidence to the XRD results that the crystalline nature enhances with the increase of milling durations.

3.4.2 TEM analysis

The bright field TEM images of MTO nanoparticles milled for 20 and 35 hrs are shown in **Figure 7(a)** and **(b)** respectively. From TEM images it was evident that the powders milled for 20 hrs does not exhibit that much distinct particles (clustering of MTO nanoparticles), but for 35 hrs milled powders a clear nanosized

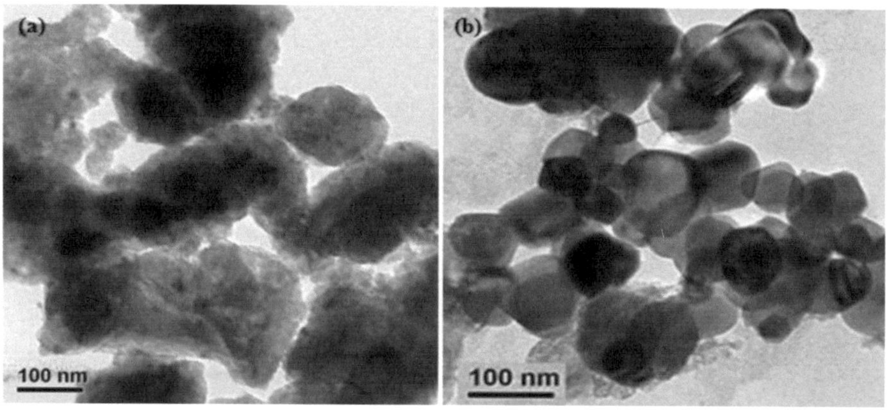

Figure 7.
Bright field-TEM micrographs of (a) 20 and (b) 35 hrs milled MTO powder.

particle is observed. The size of the parent oxides is about 2 μm. As the milling time increases the size of the initial particles decreases and for 35 hrs milled nanocrystalline MTO powders the average particle size is found to be around 60–120 nm. This crystallite size is nearly consistent with the calculated data by Williamson-Hall plot method.

The selected area electron diffraction (SAED) pattern of the 35 hrs milled powders is shown in **Figure 8(a)**. The SAED ring pattern indicates that the phase of MTO- nanoparticles was polycrystalline in structure and the distance between crystalline planes (i.e., inter planer spacing or d-spacing) was consistent with the standard pattern for a spinel MTO crystal structure. To analyze the lattice fringes for the 35 hrs milled powders, high resolution- TEM (HR-TEM) study was carried out, (as shown in **Figure 8(b)**). However, the clear lattice fringes showed that each particle has single crystalline structure. For 35 hrs milled powders, the distance between crystalline planes is evaluated and is found to 2.968 Å, which indicates the preferable crystal growth plane is (220). Thus, it could be concluded that during the high energy mechanical alloying a solid-state reaction between MgO and TiO_2 took place at room temperature.

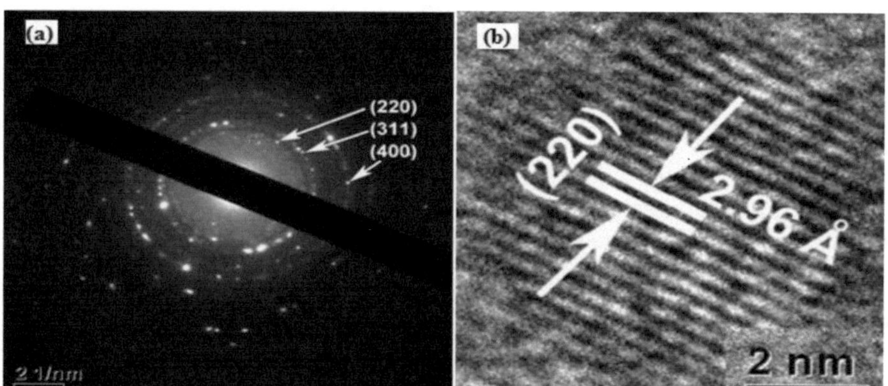

Figure 8.
(a) The selected area electron diffraction (SAED) pattern and (b) HR-TEM for 35 hrs milled powder (Adapted with permission from Bhuyan et al., 2020, @ Springer [19]).

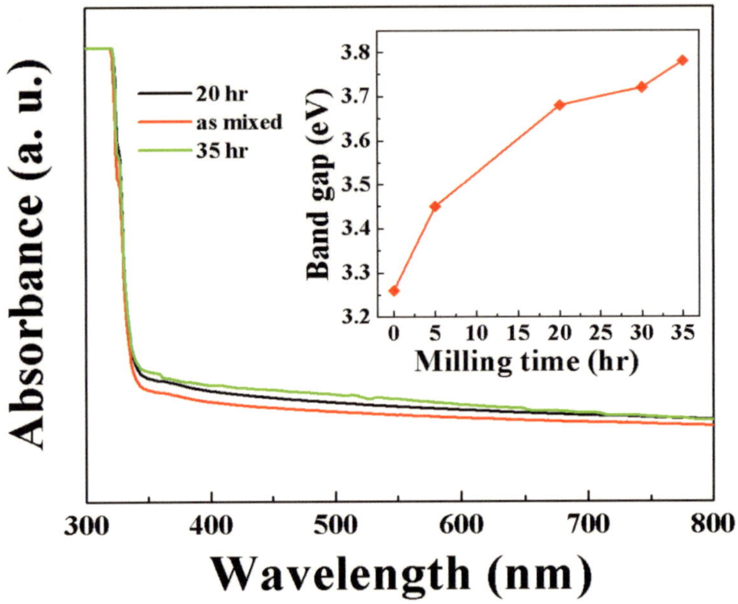

Figure 9.
Room temperature UV – Visible spectra of pure and milled MTO powders, ***inset:*** *Variation of bandgap with milling time (adapted with permission from Bhuyan et al., 2020, @ Springer [19]).*

3.4.3 Optical properties

Room temperature UV – Visible spectra were taken for all the as-prepared samples in order to see the effect of milling duration on the optical band gap of the mechanically alloyed MTO nano-powders, and are illustrated in **Figure 9**. A strong absorption peak at around 356 nm is observed for un-milled MTO powders, while with increase in milling duration, the peak slightly shifted to 352 nm for 35 hrs milled powders. It shows that there is a clear sign of blue-shift in the absorption peak with decrease in average crystallite size. This indicates that with decrease in particle size the band-gap increases. However, the enhanced absorption in mechanically alloyed MTO nanoparticles can be attributed to a large surface to volume ratio and enhanced oscillator strength with decrease in average particle size.

Tauc relation [40], is employed to estimate the optical band gap of all the milled samples. According to this relation, $\alpha h\upsilon = \beta\ (h\upsilon - E_g)^{\ n}$, where, $h\upsilon$ is the photon energy, β is a constant which measures the crystalline order of the samples and $n = 1/2$ for direct bandgap structure (As MTO belongs to cubic structures and exhibiting direct band gap). The variation of bandgap with milling time is plotted and shown in inset of **Figure 9**. The plot indicates that with increasing milling time from 0 to 35 hrs the bandgap enhanced from 3.26 eV - 3.78 eV. This result is consistent with the previously reported results [40–42]. The optical bandgap (E_g) of all the milled powders are determined by the extrapolation of the best linear fit between $(\alpha h\upsilon)^2$ and $h\upsilon$ to intercept the $h\upsilon$ axis ($\alpha = 0$), (taken along x-axis). The figure shows the dependence of the absorption coefficients $(\alpha h\upsilon)^2$ with photon energy. The position and slope of the optical absorption edge makes this material as a suitable UV light absorber.

4. Conclusions

Mg_2TiO_4 nanocrystalline with spinel structure were synthesized from high purity MgO and TiO_2via high energy ball milling techniques. The impact of milling time on particle size, crystal structure and the microstructure of mechanically derived Mg_2TiO_4 nanocomposite powders were investigated using X-ray diffraction, scanning electron microscopy and transmission electron microscopy (TEM) techniques respectively. Williamson-Hall method was employed to understand the origin of the broadening of the X-ray diffraction peaks. It was confirmed that the W-H method is a more accurate method as compared to the Scherrer method for the estimation of crystallite size of the Mg_2TiO_4 nanocomposite materials. Further, the thermal decomposition behavior of the milled powders was examined by a thermo-gravimetric analyzer (TGA) in argon atmosphere. The UV–visible spectra showed strong bandgap absorption at ~356 nm and with an increase of milling times from 0 to 35 hrs, there is an increase of the band-gap from 3.68-3.78 eV. The Mg_2TiO_4 nano- powders synthesized via mechanical alloying method showed promising optical properties which is suitable for commercial optoelectronic applications. Also, the high optical absorption edge makes this material as suitable UV light absorber. Moreover, Mg_2TiO_4 is an excellent microwave dielectric material having wide band gap and high refractive index and practically useful for various optical and electronic applications.

Acknowledgements

The author RKB acknowledges Prof. Sukant Kumar Tripathy, Department of Physics, Berhampur University, Berhampur, Odisha, India for his suggestion and discussion for the betterment of the manuscript.

Author details

Ranjan Kumar Bhuyan[1*], Bhagban Kisan[2], Santosh Kumar Parida[3], Soumya Patra[4] and Sunil Kumar[5]

1 Centre of Excellence in Nanoscience and Technology for the Development of Sensor, Department of Physics, Berhampur University, Berhampur - 760007, India

2 P.G. Department of Physics, Utkal University, Vanivihar Bhubaneswar - 751004, India

3 Department of Physics, ITER, Siksha 'O' Anusandhan Deemed to be University, Bhubaneswar - 751030, India

4 School of Applied Physics, KIIT, Bhubaneswar - 751024, India

5 Department of Physics, Ramjas College, University of Delhi - 110007, India

*Address all correspondence to: ranjanphysics.bhuyan8@gmail.com

References

[1] Rabha, Susmita., Dobbidi, Pamu.; Enhanced microwave dielectric and electrical properties of Zn substituted Mg_2TiO_4 ceramics for RF/ microwave applications, *J. Mater. Sci. Mater. in Electronics,* 60, **(2019)** 392.

[2] Yu, Hongta. Luo, Ting., He, Lie., Liu Jing song.; Effect of ZnO on Mg_2TiO_4 – $MgTiO_3$- $CaTiO_3$ microwave dielectric ceramics prepared by reaction sintering route, *Adv. in Appl. Ceram.*, 118, **(2018)** 1.

[3] Kumar, T. S.; Gogoi P.; Perumal, A.; Sharma P.; Dobbidi, Pamu.; Effect of cobalt doping on the structural, microstructure and microwave dielectric properties of $MgTiO_3$ ceramics prepared by semi alkoxide precursor method, *J. Am. Ceram. Soc.*, 97, **(2014)** 1054.

[4] Bhuyan, R. K.; Thatikonda, S. K.; Pamu, D.; Liquid phase effect of Bi_2O_3 additive on densification, microstructure and microwave dielectric properties of Mg_2TiO_4 ceramics. *Ferroelectrics*, 516 (2017) 173.

[5] Li, B. J.; Wang, S. Y.; Liao, Y. H.; Chen, Y. B.; Dielectric properties and crystal structure of $(Mg_{1-x}Co_x)_2(Ti_{0.95}Sn_{0.05})O_4$ ceramics. *J. Ceram. Soc. Jpn.*; 122, **(2014)** 955.

[6] Bhuyan, R. K.; Thatikonda, S. K.; Goswami, D. James, A. R.; Perumal, A.; Pamu, D.; Enhanced densification and microwave dielectric properties of Mg_2TiO_4 ceramics added with CeO_2 nanoparticles.*; Mater. Sci. Eng. B.;*178 (2013) 178.

[7] Bhuyan, R. K.; Thatikonda, S. K.; Goswami, D. James, A. R.; Pamu, D., Liquid phase effect of La_2O_3 and V_2O_5 on microwave dielectric properties of Mg_2TiO_4 ceramics. *J. Electroceramics,* 31 (2013) 48.

[8] Huang, C. L.; Chen, J. Y.; Low-loss microwave dielectric ceramics using $(Mg_{1-x}Mn_x)_2TiO_4$ (x =0.02-0.1) solid solutions. *J. Am. Ceram. Soc.*, 92 (2009) 675.

[9] Huang, C. L.; Ho. C. En.; Microwave dielectric properties of $(Mg_{1-x}Ni_x)_2TiO_4$ (x =0.02-0.1) ceramics. *J. Am. Ceram. Soc.*, 7(2010) E163.

[10] Cava, R. J.*;* Dielectric materials for applications in microwave communications. *J. Mater. Chem.*, 11, **(2001)** 54.

[11] Huang, C. L.; Wange, J. J.; Chang, Y.P. Dielectric Properties of Low Loss (1-x) $(Mg_{0.95}Zn_{0.05})TiO_3$-$x$$SrTiO_3$ Ceramic System at Microwave Frequency, *J. Am. Ceram. Soc.*, 90 **(2007)** 858.

[12] Huang, C. L.; Chen, J. Y.; Li, B. J., Effect of $CaTiO_3$ addition on microwave dielectric of $Mg_2(Ti_{0.95}Sn_{0.05})O_4$ ceramics, *J. Alloys Compd.*, 509 **(2011)** 4247.

[13] Chen, Y. B., $La(Mg_{1-x}Zn_x)_{1/2}Ti_{1/2}O_3$ ceramics at microwave frequencies. *J. Alloys Compd.*, 509 (2011) 1050.

[14] Belous, A.; Ovchar, O.; Durilin, D.; Krzmanc, M. M.; Valant, M.; Suvorov, D. High –Q microwave dielectric materials based on the spinel Mg_2TiO_4. *J. Am. Ceram. Soc.*, 89 **(2006)**, 3441.

[15] A. Wechsler, R. B. Von Dreele, Structure refinements of Mg2TiO4, MgTiO3 and MgTi2O5 by time-of-flight neutron powder diffraction. *Acta Cryst.*, B45 (1989) 542.

[16] Kimmel G., Zabicky J., International centre for diffraction data, *Advances in X-ray analysis*. 42 **(2000)**, 238.

[17] Pfaff, G.; Peroxide route for synthesis of magnesium titanate powders of various compositions. *Ceram. Int.*, 20 **(1994)**111.

[18] M. Kamruddin, P. K. Ajit Kumar, R. Nithya, A. K. Tyagi, B. Raj, Synthesis of nanocrystalline ceria by thermal decomposition and soft-chemistry methods. *Scripta Materialia*, 50 (**2004**) 417.

[19] Bhuyan R. K, Mahapatra R. K, Nath G., Sahoo, B. K., Das Debadutta and D. Pamu, Influence of high energy ball milling on structural, microstructural and optical properties of Mg_2TiO_4 nanoparticles. *J. Mater. Sci. Mater. Electro.*, 31 (**2020**) 628.

[20] Bhuyan, R. K.; Thatikonda, S. K.; Pamu, D.; Structural and microwave dielectric properties of Mg_2TiO_4 ceramics synthesized by mechanical method. *Int. J. Appl. Ceram. Technol.*; 516 (**2013**) E18.

[21] Cheng, L., Peng, L., Qu, S. X., Cheng, L., Zhang, H. Wu., Microwave dielectric properties of Mg_2TiO_4 ceramics synthesized via high energy ball milling method, *J. Alloy. Compd.*, 623 (**2015**) 238.

[22] Filipovic, S., Obradovic, N., Pavlovic V. B., Markovic, S., Mitric, M., Ristic, M. M., Influence of mechanical activation on microstructure and crystal structure of sintered MgO-TiO_2 system, *Science of sintering*, 42 (**2010**) 143.

[23] H. Yang, Y. Hu, A. Tang, S. Jin, G. Qiu, Synthesis of tin oxide nanoparticles by mechanical reaction. *J. Alloy. Compd.*, 363 (**2004**) 271.

[24] K. T. Paul, S. K. Satpathy, I. Manna, K. K. Chakraborty, G. B. Nando, Preparation and Characterization of Nano structured Materials from Fly Ash: A Waste from Thermal Power Stations, by High Energy Ball Milling. *Nanoscale Res. Lett.*, 2 (**2007**) 397.

[25] L. B. Kong, J. Ma, W. Zhu, O. K. Tan, Preparation and characterization of PLZT ceramics using high-energy ball milling. *J. Alloys Compd.*, 322 (**2001**) 290.

[26] C. Suryanarayana, *Mechanical Alloying and Milling*, Marcel Dekker, (**2004**).

[27] Rietveld, H.M.: Line profiles of neutron powder diffraction peaks for structure refinement. *Acta. Crystallogr.* 22 (**1967**) 151.

[28] Balzar, D., Ledbetter, H.J., Voigt function modeling in Fourier analysis of size and strain broadened X-ray diffraction peaks. *Appl. Crystallogr*. 26, (**1993**)97.

[29] Warren, B.E., Averbach, B. L., The effect of cold-work distorsion on X- ray pattems. *J. Appl. Phys.* 21 (**1950**) 595.

[30] Suryanarayana, C., Norton, M.G., X-ray diffraction: a practical approach. Plenum Press Publishing, New York (**1998**).

[31] Soni P. R, Mechanical alloying: Fundamental and applications, Cambridge International Science Publishing, UK (**2001**).

[32] Suryanarayana C, Mechanical alloying and milling, *Prog. Mater. Sci*. 46 (**2001**) 1-184.

[33] Koch C. C, Nanostructured materials: Processing, properties and potential applications, Noyes Publications, New York (**2002**).

[34] Ding J, Li Y, Chen L. F, Deng C.R, Shi Y, Chow Y. S, Gang T. B, Microstructure and soft magnetic properties of nanocrystalline Fe–Si powders. J. Alloys Compd. 314 (**2001**) 262.

[35] G. K. Williamson, W. H. Hall, X-ray line broadening from filed Aluminium and wolfram, *Acta Mater.*, 1 (**1953**) 22.

[36] Biju, V., Sugathan, N., Vrinda, V., Salini, S.L.: Estimation of nano crystalline silver from X-ray diffraction

line broadening. *J. Mater. Sci*. 43 (**2008**) 1175.

[37] Cullity, B.D., Stock, S.R.: Elements of X-ray diffraction, 3rd edn. Prentice Hall, New Jersey (**2001**).

[38] J. Xu, G.S. Collins, L. Peng, M. Atzmon, Deformation-assisted decomposition of unstable $Fe_{50}Cu_{50}$ solid solution during low-energy ball milling. *Acta Mater*., 47 (**1999**)1241.

[39] Bhuyan, R. K.; Pamu, D.; Sahoo B.K.; Sarangi, A. K.; Structural and thermal study of Mg_2TiO_4 nanoparticles synthesized by mechanical alloying method. *Micro and nanosystem,* 12 (**2020**) 87.

[40] J. Stade, D. Hahn, R. Dittmann, *Journal of Luminescence* 8 (**1974**) 308-317.

[41] H. Kominami, M. Tanaka, K. Hara, Y. Nakanishi, Y. Hatanaka, *Physica Status Solidi C* 3 (**2006**) 2758-2761.

[42] A. Golubovic, M. Radovic, *Journal of Serbian Chemical Society*, 76 (**2011**) 1561-1566.

Chapter 8

Magnesium Metal Matrix Composites and Their Applications

Aqeel Abbas, Veeramanikandan Rajagopal and Song-Jeng Huang

Abstract

Magnesium is one of the lightest structural metals have the capability to replace the conventional alloys for mass saving applications and provides higher strength and stiffness. Additionally, it also has the ability to absorb the hydrogen in the form of hydrides and can be used as a future source of energy carrier. The theoretical hydrogen capacity of 7.6 wt% makes it more suitable for future energy sources but needs to reduce the working sorption temperature. Moreover, magnesium is the primary source of the body and has strength equal to the bone, making it more suitable for biomedical applications and higher biocompatibility. Some challenges of magnesium-based metal matrix composites are still encountering structural applications, hydrogen energy storage, and biomedical applications due to manufacturing methodologies and proper materials selection to get required results.

Keywords: magnesium-based metal matrix composites, structural applications, hydrogen energy storage, biomedical applications of magnesium, automotive applications

1. Introduction

In recent decades, the modern world focuses on the research of lightweight metallic materials with high strength for potential applications [1]. Over the past few years, different steel materials and its alloys have been used for the applications such as household constructions and automotive manufacturing industries due to its high strength properties. Even though steel alloys have high strength, many of the light metal alloys systems also have sufficient strength to make sure of their usage in particular applications [2].

Aluminum and magnesium alloys are widely used in industries due to low mass density in recent days. The higher strength to mass ratio makes them as most attractive materials where reducing weight is one of the most significant importance, such as electronic frame production, sports goods, spacecraft machinery production, and ground transports [2]. Even the manufacturing industries using a different kind of light metals with strategic significance are also using magnesium due to its lowest density of 1.74 g/cm^3. The magnesium density is two-third of aluminum, one-fourth of zinc, and one-fifth of steel. Among the magnesium alloys, the AZ series are widely used because of their superior properties such as excellent castability, damping capacity, and higher machinability. In general, the magnesium alloys have low corrosion resistance and low mechanical strength, especially at elevated temperatures [3]. The necessity of lightweight materials for use in challenging applications has spurred widespread efforts to develop magnesium- metal matrix composites and cost-effective fabrication technologies. So, our research team focused on the metal matrix composites system to reach unattainable properties by the single material system.

Apart from several applications, magnesium and its alloys are considered as one of the prominence energy storage materials which can store hydrogen gas in the form of magnesium hydride [4]. Magnesium has a maximum storage capability of 7.6 wt.% as theoretical with excellent thermodynamic reversibility among the current storage materials. However, the processing temperature (300 °C) of magnesium hydrides is too high to reach the target of the department of energy of the united states of America. These poor properties and other issues have limited their commercial implementations that need to be optimized [5].

Magnesium alloys with different additive materials are quite attractive content for scientific investigations. The prepared materials' performance depends on the composition, fabrication methodology, processing techniques, and properties will be varying with materials that added to the magnesium [6]. To improve the material properties, several types of additives such as SiC, Al2O3, carbon allotropes, B4C, TiC, and transition metals (particles/whiskers) have been fabricated using liquid state technique (stir casting, squeeze casting, centrifugal casting) and semi-solid-state techniques (chemical vapor deposition and physical vapor deposition) to attain the superior properties for various applications [7–10]. The above-mentioned techniques are selected depending upon what kind of reinforcement distribution is required in matrix and how much cost can be effective. The secondary processing technique (heat treatment and plastic deformations) influences composites' micro-structures and specific behaviors and alloy materials. The secondary processing techniques are used to refine the microstructure to enhance the ductility and strength of the composites.

By investigating these materials, the characteristics of recrystallization in the magnesium and phase transformations could be recognized towards the applications [11, 12]. We are hopeful that our research studies on magnesium with different additive materials contribute the breakthrough knowledge in the field of mechanical behaviors and energy storage technology, which are viable to future applications.

2. Manufacturing methodologies to fabricate the metal matrix composites

The metal matrix composites (MMCs) composed of low-density magnesium reinforced with fibers or particles. MMCs offer high strength and high stiffness operating at higher temperatures and possess excellent wear-resistant properties as the properties are tailored according to required properties. However, MMCs have many disadvantages than metallic alloys, such as higher manufacturing costs for better performance and low ductility. This materials group has become attractive for the use in construction and functional applications. The demand for MMCs materials has been increased and is expressed in **Figure 1**.

The manufacturing methodologies are categorized into following

- Solid-state processing

- Liquid state processing
- In-situ processing

Each process is further categorized and express in the following chart.

2.1 Solid-state processing

The principal manufacturing methodologies included in solid-state processing are physical vapor deposition, powder blending and consolidation, and additive manufacturing.

2.1.1 Physical vapor deposition

The fibers are continuously passed over the substrate of the metal with high partial pressure. The vapors produced are inserted in the process and condensed to make the coating on the surface. The deposition rate is 5-10 μm per minute. The coating fibers are consolidated by hot isostatic pressing. Sputtering is a physical vapor deposition method in which argon ions are used for bombarding catholically connected target to deposit the coating.

2.1.2 Powder blending and consolidation

The magnesium powder is mixed with reinforcement to get uniformity under a neutral environment. The blending steps, along with powder size, control the mechanical properties of composites. The uniform and homogeneous mixture is compacted under controlled temperature. The oxides are supposed to be produced on the surface as magnesium is highly reactive to oxygen.

2.1.3 Additive manufacturing

The reinforcements are uniformly distributed and dispersed in the matrix using 3D manufacturing techniques. The complex geometry and graded density

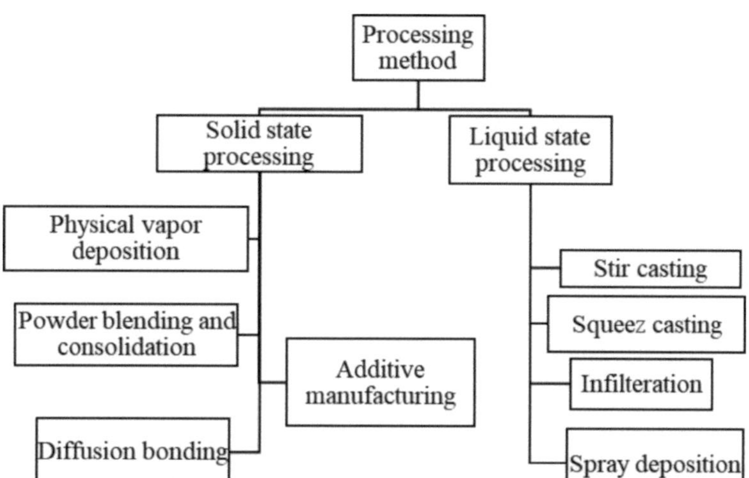

Figure 1.
Manufacturing methodologies for MMCs.

composites can be developed. The protection of magnesium powder from oxides is very critical in this method.

2.1.4 Diffusion bonding

It is a very important solid-state technique to combine two metals. The cleaned surfaces are pressed at elevated temperature leading to bonding of the metals.

2.2 Liquid state processing

The liquid state processing is cost-effective and complex geometries can be achieved with good interfacial bonding and excellent dispersion. The excellent bonding and uniform dispersion of reinforcement lead to enhancement in mechanical properties. Stir casting, squeeze casting, and infiltration are the prominent liquid state processing techniques.

2.2.1 Stir casting

The particle reinforcement is mixed into a molten matrix using rotating impellers and then solidified to room temperature. The problem raised during stir casting is uniform distribution and sediments in the molten alloy. The 30% of the particles in size of 5–100 micrometers can be incorporated in the metal alloy.

2.2.2 Squeeze casting

The molten metals are poured into the closed die and pressed under some constant pressure. The heat is transferred from molten metal to die. With the high pressure of dies, grain refinement happens, and castings have very little porosity and lead to the composites' increased strength.

2.2.3 Infiltration process

The molten metal alloy is unfiltered into the porous form of reinforcement under the pressurized gas to apply the pressure. The pressure causes the penetration and dispersion of molten metal into reinforcement, which have a volume fraction of 10–70% depending upon porosity level. Some binding agents are required to maintain the integrity and shape of porous form. The infiltration processes are classified into the following three categories

- Gas Pressure Infiltration
- Spray Deposition
- Chemical Vapor Deposition (CVD)

The gas is applied in the gas infiltration process to penetrate the molten metal into porous reinforcement. The reinforcements of porosity 5–10% are sprayed on the surface of metal alloy in the spray deposition process. The sprayed particles are consolidated by further processing. The particle/fibers spacing impacts the volume fraction and distribution.

2.2.4 Spray deposition

The spray deposition is a technique in which droplets of molten metal or continuous feeding of cold metal is spread over rapid heat injection zone. The process is adopted for bulk production by directing the atomized stream of droplets on the substrate. The adaptation of this technique for the building of MMCs by injecting ceramic powder into spray has commercially succeeded. The droplet/powder velocities are typically 20-40 m/s, and the thin layer of semisolid reinforcement is present on the surface of the ingot. The MMCs produced by spray deposition often exhibit inhomogeneous distribution, and the ceramic layer is normal to growth direction, which leads to hydrodynamic instabilities in powder injection. The porosity in MMCs in the as-sprayed state is typically 5–10%.

In the chemical vapor deposition process, the vapors react/decompose and form a coating on another substrate. The process is carried out at elevated temperatures. The in-situ processing is a unique process in which chemical reaction results in the development of reinforcement in MMCs. The reinforcement can be formed from precipitations of the liquid of solid. The method provides the thermal compatibility between matrix and reinforcement, and surfaces are free of contaminations.

3. Secondary processes

The processes performed after the manufacturing of the composites to enhance the strength and modify the microstructure are often referred to as a secondary process. The most common secondary processes are Heat treatment and Plastic deformation.

3.1 Heat treatment

The processes of heat treatments are performed to dissolve some phases and to generate some new phases. There are different processes and methods to perform heat treatments, which are explained in **Table 1**. The heat treatment processes increase the corrosion resistance and strength of the composites. The heat treatments are classified into the following categories

- Solution heat treatment
- Homogenization
- Aging

In solution heat treatments, the casted alloys/composites are heated to a specific temperature for a particular period of time to develop the desired constituent in the matrix. The billets are then quenched suddenly to sustain the phases develop.

Homogenization heat treatments are performed to enhance the workability of cast billets. The brittle nonequilibrium phases are dissolved, and a homogenous microstructure is developed. The homogenization of the billets is performed at the recrystallization temperature of the materials.

Aging is referred to as the process of heat treatment in which precipitates developed during homogenization are decomposed at some elevated temperature. The aging leads

Suffix	Heat treatment conditions
T1	Cooled from elevated temperature and naturally aged
T2	Cooled from elevated temperature, cold worked and naturally aged
T3	Cooled from elevated temperature, quenched, cold worked, and naturally aged
T4	Solutionized, quenched, and naturally aged
T5	Rapidly cooled and artificially aged
T6	Solutionized, quenched, and artificially aged
T7	Solution heat treatment, quenched and overaged
T8	Solution heat treatment, cold worked and artificially aged
T9	Solution heat treatment, artificially aged cold worked
T10	Cooled from elevated temperature (recrystallization temperature) cold worked and artificially aged

Table 1.
Types of heat treatment employed to MMC for microstructure modification.

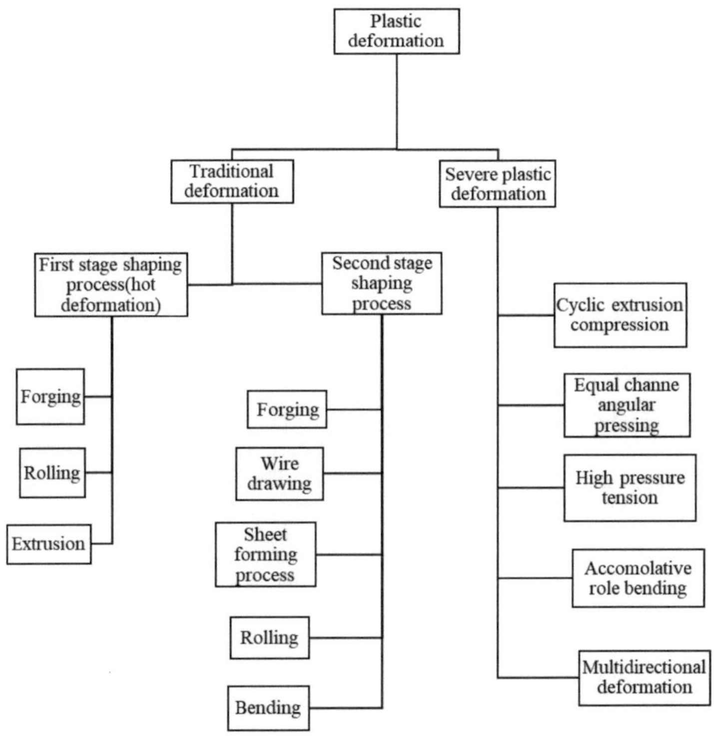

Figure 2.
Types of plastic deformation used for MMCs.

to an increase in yield strength and hardness of the alloy/composite. The aging process may be natural (at room temperature) or artificial aging (at elevated temperature).

Some terms have been introduced by the aluminum association for aluminum and its alloys and are explained in **Table 1**.

3.2 Plastic deformation

The metal matrix composites are deformed into different shapes, and the microstructure is changed to increase the strength of the composites. The plastic deformation can be categorized into traditional deformation and severe plastic deformation. The deformation could be hot or cold deformation. The shape of the composite is changed, and mechanical properties are moderately enhanced in traditional deformation, but severe plastic deformation alters the microstructure, and mechanical properties are exceptionally increased. The composition of the microstructure is changed, but the shape of the composite remains the same. The type of plastic deformation is given in the following flow chart **Figure 2**.

4. Structural applications

The magnesium is the third most commonly used structural material after aluminum and steel and is significantly used in automotive, power tools, aerospace, and 3C (computer, communication, and consumer products). Currently, magnesium application in automotive includes transfer case, radiator support, instrument penal beam, and steering components. Magnesium with minimal density is used in mass saving applications to replace aluminum and steel, but magnesium alloys' strength is low. The hard and tough ceramics are added in magnesium alloys using proper manufacturing techniques to improve the strength for structural applications. The extrusion of the metal matrix composites provide strength comparable to aluminum alloys. AZ, ZK and AM series of the magnesium alloys are very common to fabricate the metal matrix composites. The magnesium based metal matrix composites are developed because the wrought magnesium alloys are very less formable at room temperature than aluminum alloys due to hexagonal closed packed crystal structure, although ductility appears reasonable [13].

The materials selection for structural applications is very complex in which component geometries, loading conditions, manufacturing process, materials properties, and the cost is very important. The bending mode is the primary loading condition in automotive structures. Thus, the calculation should be based on bending stiffness and strength. The thickness and mass ratio of magnesium alloy (AZ91) should be 1.67 and 0.39 times the steel to get similar bending strength. AZ91 AZ31 and AZ61 are the most commonly used magnesium alloy in the automobile. The addition of proper reinforcement in AZ series of the magnesium alloys have improved the strength and ductility. AZ61 and AZ80 have higher strength but less extrudability. The higher strength alloy ZK60 is designed for racing cars, bicycle parts, but extrusion speed is very low. The implementation of severe plastic deformation on magnesium alloys have improved the mechanical strength drastically. The new magnesium composites are being developed with a higher extrusion rate and maintaining good mechanical properties such as AM30 (higher strength applications) and ZE20 (higher ductility applications) reinforced with Al2O3, WS2, TiC, and SiC. The presence of aluminum contents in magnesium-based composites improves the strength, hardness, and corrosion resistance but reduces the ductility. The aluminum contents within range of 5–6% yield optimal strength and ductility. Zinc is next to the aluminum alloying element to improve the corrosion resistance but reduces the ductility. Manganese does not affect the tensile strength but increases the yield strength. It is 0.4% recommended by ASTM specification B93-94a to improve the corrosion resistance [14, 15].

The small addition of ceramics (Al2O3, TiC, B4C, SiC, WS2, MoS2 etc.) in magnesium improves the ductility and likely reduces the grain size and weakens the texture. The combined addition of zinc and cerium enhances the strength and ductility closer to aluminum. The addition of lithium, zirconium, and cerium has improved the formability and strength of the composites.

4.1 Aerospace applications

Magnesium alloys were extensively used in aerospace in world war I and world war II. The united states air force's bombers B-36 and B52 contain a large amount of magnesium in casting, forging, and extrusion. The B-36 uses magnesium alloys ranging from 66 lb. to 12200 lbs. in different forms like sheets, castings, and extrusion. The Boeing 727 aeroplanes contain 12 parts, including control surfaces, wheels, engine gearbox, structural items, door frames, and edge flaps are made of magnesium alloys. Soviet air crafty industry, including the TU-95MS plane and TU-134, are using magnesium alloys up to 780Kg at different locations [13]. The International Air Transport Association (IATA) legislation has limited the magnesium application due to magnesium alloys' higher corrosion performance. Some of the commonly used alloys and their applications are mentioned in **Table 2**. It is expected that magnesium will be a major element in a future world in structural materials in aerospace.

4.2 Automotive applications

The first application of magnesium in automotive was in engines of racing cars. England used commercial magnesium applications such as in crankcases of busses and tractors' transmission housing in the 1930s. The demand for magnesium applications in automotive was increased in world war I and II. The major parts which use magnesium are given in **Table 3**. The major problems in applications of magnesium in automotive are limited thermal conductivity, and magnesium cannot perform well between hot and cold temperatures. The magnesium strength is increased by doping the different reinforcement and deformation process to get the required strength and ductility [16, 17].

4.3 Electronics applications

Magnesium and its composites are being opted for by the electronics industry owing to its mass reduction. These materials are most commonly used in audio/video players, computers, mobile phones, radar detectors, and many more. Magnesium makes the electronics industry smarter and lighter due to high strength and durable to protect highly sensitive technology. Alloys and composites should be developed to improve the mechanical properties by precipitation hardening [18].

Alloy	Applications	Alloy	Application
ZE41	Sikorsky UH60 Family	WE43A	Sikorsky S92 main transmission
AZ92A	Boeing 737, 747, 757 and 767	WE43	Pratt & Whitney F119 auxiliary casing
ZE41	Pratt & Whitne Canada	ZRE1	Rolls-Royce tray

Table 2.
Common applications of magnesium in aerospace.

VW engine	Halibrand racing wheels	Buick car of the future	GM production wheels
Alfa Romeo seat	Ford radiator support	BMW engine block	BMW door inner
Mercedes transmission case	Interior	Body	Chassis

Table 3.
Common applications of magnesium in automotive.

5. Biomedical applications

Magnesium is present in the form of the mineral is 65% in bones and teeth, and the remaining 35% is present in body fluids and tissues. The biocompatibility and biodegradability make magnesium more suitable for biomedical applications. The researchers are paying much attention to influence the interactive mechanism of biodegradable materials. The magnesium strength has approached the bone strength and is used in orthopedic implants. The magnesium can also be used in other orthopedic surgeries such as screws, plates, and fasteners. The magnesium shows nominal changes in blood composition with six months of implantation without damage to the liver and kidneys [19].

The magnesium is a major alloying element used in biodegradable implant in vivo and in vitro conditions. It has excellent biocompatibility and is the fourth most abundant element present in the human body, and it is an essential nutrient element. Magnesium has a high corrosion rate and releases H_2 gas in human body fluid. Modern techniques have been adopted to control the corrosion rate and heal the fractured tissues without the need for secondary surgery to remove the implants. The fraction of the component must be selected as per biocompatibility to avoid toxicity. Thus, extreme concerns must be taken to choose the reinforcement with magnesium to control biomechanical properties and corrosion rate under biological conditions. The selection of bioceramics is very important as they may cause some severe body fluid issues, and some are mentioned in **Table 4** [20, 21]. The pure magnesium implant is degraded much earlier than the tissue heals. Therefore, mechanical and degradation properties are intended to increase.

The addition of alloying elements in pure magnesium improves the grain refinement and strengthen the composites. The most commonly used bioceramics and their effects are presented in **Table 5**. These ceramics are added in magnesium and its alloys to fabricate the magnesium-based composites with excellent strength and biodegradable properties. The biocompatibility of composite in the biological environment is based on strengthening ability degradation and toxicity [22].

The alloying elements can be classified as toxic, nutrient, and allergic, which are present in the human body. The bioinert and bioactive ceramics are significant reinforcements to achieve required biocompatibility and strength. The properties of

Toxic	Allergic	Nutrient
Ba, Pb, Be, Th, Cd	Al, Ce, Cr, Co, Cu, La, Ni, Pr, V	Ca, Mn, Sn, Sr., Zn

Table 4.
Bioelements and their effects on biofluids.

Ceramic type	Characteristics	Applications
Alumina (Al2O3)	Biocompatible and bioinert, good strength, high hardness, the non-advocate tough membrane at the interface	Permeable coatings for stems, screws, and plates, knee prosthesis
Zirconium (ZrO2)	Elevated fracture toughness, flexural strength, bioinert, biocompatible, non-toxic	Artificial knees, bone screws and plates, etc.,
Hydroxyapatite (HAP)	Bioresorbable, bioactive and biocompatible, similar composition to the bone, good osteoconductive properties	Femoral knee, femoral hip, tibial components, the acetabular cup

Table 5.
Commonly used bioceramics and their characteristics.

common bioceramics are given in **Table 5**, and the selection of specific bioceramic is important for outstanding performance in toxicity and immunological environment.

6. Hydrogen storage applications

6.1 Role of magnesium in hydrogen storage

In the recent decade, rapid advancement in nanostructuring techniques brought a new confidence in hydrogen storage applications. The hydrogen gas has been stored in metal hydride in the new generation technology in solid-state hydrogen storage. The physical and chemical properties of materials could be fundamentally varied with nano parameters. The research focused on nano-microstructures, which can play a vital role in hydrogen storage materials [23, 24]. To attain the maximum hydrogen storage capacity, Mg-based hydride materials are considered the most promising metallic content based on materials. Even though Mg has a higher storage capacity, the magnesium hydride's practical applications are still not reached because of its high working temperature and low kinetics. Besides, the high thermostability of bulk MgH_2, such as Entropy (ΔS), Enthalpy (ΔH) with the decomposition, is 130JK-1 mol-1H2 75 kJ/mol H_2 respectively, which means the temperature needed to reach 300 °C at the equilibrium pressure of 1 bar. So, it is required to optimize the stability between magnesium and hydrogen molecules that could increase the thermodynamics and kinetics properties. Reducing the particle size (nanoscale) and adding various catalyst materials are two critical factors in enhancing hydrogenation properties. The following factors that influence the absorptions rate are the physisorption rate of hydrogen gas on the magnesium alloy's surface, dissociation capacity, and nucleation growth of magnesium to the magnesium hydride [25, 26].

6.2 Mechanism of hydrogenation in magnesium-based materials

The pure magnesium, which has a hexagonal structure reacts with hydrogen molecules reversibly to form magnesium hydride in which the parametric details are presented in **Table 6**—the reversible reaction presented in Eq. (1).

$$\text{Hydrogen storage equation: } Mg + H_2 \rightarrow MgH_2 \quad (1)$$

The crystal structure of MgH_2 transformed into tetragonal β- MgH_2. At the same time, the pressure is increased with the high hydrogen pressure at the ambient temperature. The β- MgH_2 transformed into a metastable orthorhombic γ- MgH_2 phase. The schematic illustration of the mechanisms is presented in **Figure 3**.

The following reaction mechanism is composed of five different intermetallic processes.

1. hydrogen physisorption at the surface
2. dissociation of the hydrogen molecule/chemisorption
3. surface penetration of the hydrogen into the material
4. diffusion through the hydride layer to the interface with the metallic phase
5. conversion of metal into metal hydride [29].

Magnesium (Mg)	Magnesium hydride (MgH_2)
Crystal structure: Hexagonal Space group: p63/MMC Lattices parameters: a = 3.2075 Å, c = 5.2075 Å, α = 90°, β = 90°, γ = 120° Atomic coordination: Mg (c) (1/3, 2/3, 1/4)	Crystal structure: tetragonal rutile -type Space group: p42/mnm Lattices parameters: a = 3.025 Å, c = 4.5198 Å, α = 90°, β = 90°, γ = 90° Atomic coordination: H (f) (0.30478, 0.30478, 0) Mg (a) (0, 0, 0)

Table 6.
Magnesium and magnesium hydride crystallographic data [27].

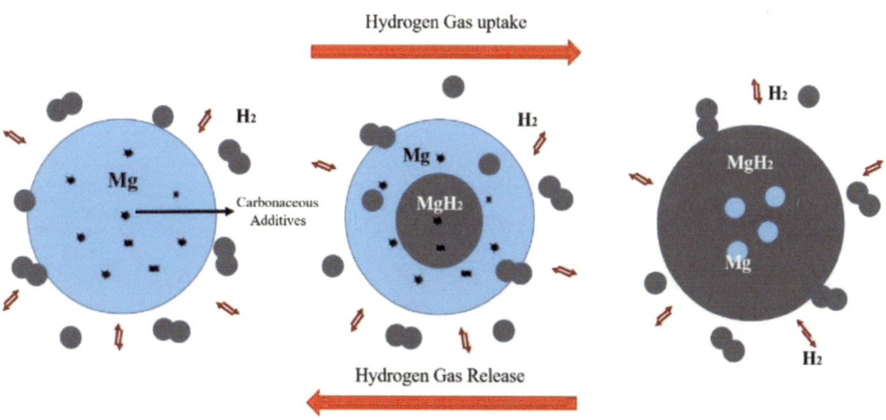

Figure 3.
The hydrogen gas ab/desorption mechanisms in magnesium [28].

6.3 Nanostructuring

Nanostructuring is also one of the important factors that can enhance the kinetics of magnesium-based materials, and it can destabilize the thermodynamics of magnesium hydride formations. The reduction of particle size into nanoscale will decrease the stability of metal hydride. The nanoparticles significantly contribute to the overall surface energy. The total energy required for dehydrogenations on metal hydrides depends on the radius of particle size r and could be as in Eq. (2) [30].

$$\Delta G(r) = \Delta G_o(r) + RT \ln \frac{P}{P_o} + \frac{3V_{M\Delta_{M \to MH2}(\gamma, r)}}{r} \tag{2}$$

6.4 Magnesium composites in hydrogen storage application

6.4.1 Mg-carbon composites in hydrogen storage

The novel electronic properties of carbon and exciting interaction between hydrogen and carbon atoms, particularly nanostructured materials, exhibit the prominent catalytic effect on the Mg as the hydrogen storage material. Initial time, graphite has been proposed as one of the anti-sticking agents in the ball milling to improve the efficiency of the process with magnesium and examined its catalytic effect on the hydrogenation characteristics. Later on, the studies investigated

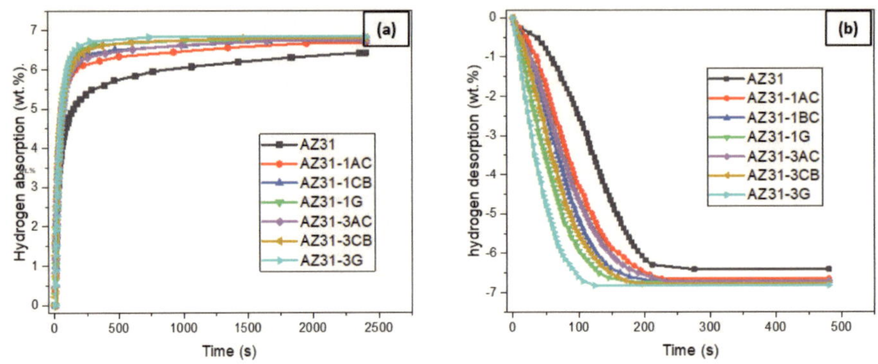

Figure 4.
Hydrogenation kinetic curves for AZ31- magnesium alloy/carbon materials, (a) Hydrogen absorption, (b) Hydrogen desorption [28].

Materials	Processing Method	Particle Size	Temperature/ Pressure	Capacity	Kinetics	Ref.
Mg-CB	Ball milling (BM)	5–15 μm	300 °C	6.0 wt.%	Abs: 40 min	[31]
Mg-Ti/ Graphene	BM 600 rpm, (10 hrs)	1.4–2 μm	300 °C/0.08 bar	4.3 wt.%	- (PCT)	[32]
Mg-GNS (Graphene Nano Sheet)	BM 450 rpm, (20 hrs)	<10 nm	300 °C	6.2 wt.%	60 min	[33]
Mg- AC (activated carbon)	BM 300 rpm, (15 hrs)	30–45 nm (crystallite)	300 °C	6.5 wt.%	7 min	[34]
AZ31 ZK60	ECAP (plastic deformation)	—	320 °C	~6.4 wt.%	12 hrs.	[35]
AZ61	ECAP(plastic Deformation)	24.09 μm (grain)	375 °C 35 atm	6.2 wt.%	Abs:2564 s Des: 156 s	[36]
AZ31/ Graphene	ECAP	4.35 ± 0.7 μm (grain)	375 °C 35 atm	6.72 wt.%	Abs:1567s Des: 229 s	[28]
	Ball milling (BM)	39.18 μm		6.83 wt.%	Abs:792 s Des: 143 s	

Table 7.
Hydrogen storage properties of Mg-alloys with different types of carbonaceous materials.

the advantages of the carbon additives such as single-walled carbon nanotubes (SWNT), activated carbon (AC), carbon black (CB), fullerene (C60), and boron nitrate (BN) nanotubes in the enhancements of practical hydrogenations properties of magnesium and its alloys.

The addition of carbon materials shows significant advantages in hydrogenation kinetics and particularly in the improvement of hydrogenation capacities. Moreover, Huang et al. investigated the hydrogenation kinetics with different carbon materials such as activated carbon (AC), carbon black (CB), and graphene nanosheets. The AZ31- magnesium alloy added with graphene sheets reached the maximum storage capacity of 6.83 wt.%. The AZ31-magnesium alloy reached its maximum capacity in less than 15 minutes and released entire hydrogen in less than 3 minutes, and its kinetic curves presented in **Figure 4**. It clearly shows that carbon and magnesium-based materials as composites (Mg alloys-carbon composites) produce the prominent catalytic effect for hydrogenation. Moreover, the results illiterate the carbon effects, which are strongly related to the unique electronic-π properties and sheet morphology of nanographene, which can make the high porosity with the high surface area. The sheet-like morphology acts as the nucleation site for the hydrogen molecules penetration into the materials and could increase the hydrogenation kinetics [28].

The theoretical capacities of different materials were calculated by the elemental molecular equation as follows

$$C = \frac{xM_{H_2}}{xM_{H_2} + xM_{Mg} + yM_{Al} + zM_{Zn} + tM_c + \ldots} \quad (3)$$

Where C is the theoretical capacity (wt.%); M_{H2} is the molecular weight of hydrogen; M_{Mg} is the molecular weight of Mg; M_{Al} is the molecular weight of Al; M_{Zn} is the molecular weight of Zn; M_C is the molecular weight of C with its corresponding weightage of the quantity in the reaction (x, y, z, and t). The hydrogen storage properties of Mg-Alloys with different carbonaceous materials were presented in **Table 7**.

7. Materials challenges

A limited number of magnesium alloys are available compared to aluminum alloys. The properties can be significantly improved by introducing micro/nano-sized particles. The reinforcements offer tremendous opportunities to improve the characteristics of the material and tailored the properties. It is very difficult to avoid oxidation while handling the melt casting. Various processes need to be optimized for magnesium alloys and Mg MMCs. New forming processes should be developed to improve formability at a certain temperature and a certain strain rate.

In the hydrogen storage applications, magnesium is one of the promising candidates even though there are many available materials. But poor kinetics and high hydriding, de-hydriding temperature limits its applications. Considerable efforts are dedicated to solving these issues by preparing composites by adding catalytic materials and nanostructuring. However, Mg-based material systems still having the parameters to be optimized. It is needed that the fabrication methods for preparing sub-nano, metric crystals with size will be considered to attain the potential targets to destabilizes the strong bondage between Mg-H. Even though the targets still stay on to be experimentally achieved.

Author details

Aqeel Abbas, Veeramanikandan Rajagopal and Song-Jeng Huang*
Department of Mechanical Engineering, National Taiwan University of Science and Technology, Taipei, Taiwan, R.O.C

*Address all correspondence to: sgjghuang@mail.ntust.edu.tw

References

[1] Abbas A, Huang S-J. ECAP effects on microstructure and mechanical behavior of annealed WS2/AZ91 metal matrix composite. J Alloys Compd 2020:155466. doi:10.1016/j.jallcom.2020.155466.

[2] Rashad M, Pan F, Liu Y, Chen X, Lin H, Pan R, et al. High temperature formability of graphene nanoplatelets-AZ31 composites fabricated by stir-casting method. J Magnes Alloy 2016;4:270-7. doi:10.1016/j.jma.2016.11.003.

[3] Sundaresan R, Froes FH. Mechanical Alloying in the Titanium-Magnesium System. Key Eng Mater 1991;29-31:199-206. doi:10.4028/www.scientific.net/KEM.29-31.199.

[4] Kichigin VI, Shein AB. The kinetics of hydrogen evolution reaction accompanied by hydrogen absorption reaction with consideration of subsurface hydrogen as an adsorbed species: Polarization curve. J Electroanal Chem 2020;873:114427. doi:10.1016/j.jelechem.2020.114427.

[5] Huang S-J, Rajagopal V, Ali AN. Influence of the ECAP and HEBM processes and the addition of Ni catalyst on the hydrogen storage properties of AZ31-x Ni (x=0,2,4) alloy. Int J Hydrogen Energy 2019;44:1047-58. doi:10.1016/j.ijhydene.2018.11.005.

[6] Huang S-J, Abbas A. Effects of tungsten disulfide on microstructure and mechanical properties of AZ91 magnesium alloy manufactured by stir casting. J Alloys Compd 2020;817:153321. doi:10.1016/j.jallcom.2019.153321.

[7] Ward-Close CM, Partridge PG, Gilmore CJ. Precipitation and hardening in titanium-magnesium and titanium-calcium alloys. Titanium'92 Sci Technol 1993:651-8.

[8] Huang S, Ali AN. Materials Science & Engineering A E ff ects of heat treatment on the microstructure and microplastic deformation behavior of SiC particles reinforced AZ61 magnesium metal matrix composite. Mater Sci Eng A 2018;711:670-82. doi:10.1016/j.msea.2017.11.020.

[9] Zhou E, Suryanarayana C, Froes FHS. Effect of premilling elemental powders on solid solubility extension of magnesium in titanium by mechanical alloying. Mater Lett 1995;23:27-31.

[10] Chen C, Cai G, Chen L, An Y, Xu G, Wang Q. Hydrogen absorption properties of a slurry system composed of a magnesium alloy powder and C6H6. J Alloys Compd 2003;356-357:409-12. doi:10.1016/S0925-8388(03)00225-1.

[11] Abbas A, Huang S-J. Investigation of severe plastic deformation effects on microstructure and mechanical properties of WS2/AZ91 magnesium metal matrix composites. Mater Sci Eng A 2020;780:139211. doi:10.1016/j.msea.2020.139211.

[12] Huot J. Nanocrystalline Metal Hydrides Obtained by Severe Plastic Deformations. Metals (Basel) 2012;2:22-40. doi:10.3390/met2010022.

[13] Luo AA. Magnesium casting technology for structural applications. J Magnes Alloy 2013;1:2-22. doi:10.1016/j.jma.2013.02.002.

[14] Friedrich H, Schumann S. Research for a "new age of magnesium" in the automotive industry. J Mater Process Technol 2001;117:276-81. doi:10.1016/S0924-0136(01)00780-4.

[15] Powell BR, Rezhets V, Balogh MP, Waldo RA. Microstructure and creep behavior in AE42 magnesium die-casting alloy. JOM 2002;54:34-8. doi:10.1007/BF02711864.

[16] Verma R, Carter JT. Quick Plastic Forming of a Decklid Inner Panel with Commercial AZ31 Magnesium Sheet, 2006. doi:10.4271/2006-01-0525.

[17] Logan S, Kizyma A, Patterson C, Rama S. Lightweight Magnesium Intensive Body Structure, 2006. doi:10.4271/2006-01-0523.

[18] Luo A, Sachdev A. Microstructure and Mechanical Properties of Magnesium-Aluminum-Manganese Cast Alloys. Int J Met 2010;4:51-9. doi:10.1007/BF03355502.

[19] Bommala VK, Krishna MG, Rao CT. Magnesium matrix composites for biomedical applications: A review. J Magnes Alloy 2019;7:72-9. doi:10.1016/j.jma.2018.11.001.

[20] Nakamura Y, Sakaki K, Kim H, Asano K, Watanuki T, Machida A. Reaction paths via a new transient phase in non-equilibrium hydrogen absorption of LaNi2Co3. Int J Hydrogen Energy 2020;45:21655-65. doi:10.1016/j.ijhydene.2020.05.237.

[21] Nakamura Y. Differences in Behavior among the Chlorides of Seven Rare Earth Elements Administered Intravenously to Rats. Fundam Appl Toxicol 1997;37:106-16. doi:10.1006/faat.1997.2322.

[22] Radha R, Sreekanth D. Insight of magnesium alloys and composites for orthopedic implant applications – a review. J Magnes Alloy 2017;5:286-312. doi:10.1016/j.jma.2017.08.003.

[23] Schneemann A, White JL, Kang S, Jeong S, Wan LF, Cho ES, et al. Nanostructured Metal Hydrides for Hydrogen Storage 2018. doi:10.1021/acs.chemrev.8b00313.

[24] Huang S, Rajagopal V, Ali AN. ScienceDirect Influence of the ECAP and HEBM processes and the addition of Ni catalyst on the hydrogen storage properties of AZ31-x Ni (x ¼ 0 , 2 , 4) alloy. Int J Hydrogen Energy 2018;44:1047-58. doi:10.1016/j.ijhydene.2018.11.005.

[25] Jia Y, Sun C, Shen S, Zou J, Mao SS, Yao X. Combination of nanosizing and interfacial effect: Future perspective for designing Mg-based nanomaterials for hydrogen storage. Renew Sustain Energy Rev 2015;44:289-303. doi:10.1016/j.rser.2014.12.032.

[26] Bardhan R, Ruminski AM, Brand A, Urban JJ. Environmental Science Magnesium nanocrystal-polymer composites : A new platform for designer hydrogen storage materials 2011:4882-95. doi:10.1039/c1ee02258j.

[27] Shao H, Xin G, Zheng J, Li X, Akiba E. Nanotechnology in Mg-based materials for hydrogen storage. Nano Energy 2012;1:590-601. doi:10.1016/j.nanoen.2012.05.005.

[28] Huang S, Rajagopal V, Chen YL, Chiu Y-H. ScienceDirect Improving the hydrogenation properties of AZ31- Mg alloys with different carbonaceous additives by high energy ball milling (HEBM) and equal channel angular pressing (ECAP) *. Int J Hydrogen Energy 2019;45:22291-301. doi:10.1016/j.ijhydene.2019.10.032.

[29] Martin M, G C, Borkhart C, Fromm E. Absorption and desorption kinetics of hydrogen storage alloys 1996;238:193-201.

[30] Radtke G, Dresselhaus M, Chen G, Be V. Size effects on the hydrogen storage properties of nanostructured metal hydrides : A review 2007:637-63. doi:10.1002/er.

[31] Spassov T, Zlatanova Z, Spassova M, Todorova S. Hydrogen sorption properties of ball-milled Mg-C nanocomposites. Int J Hydrogen Energy 2010;35:10396-403. doi:10.1016/j.ijhydene.2010.07.123.

[32] Ye W, Chi Q, Zhou H, Gao P. ScienceDirect Ball-milling preparation of titanium / graphene composites and its enhanced hydrogen storage ability. Int J Hydrogen Energy 2018;43:19164-73. doi:10.1016/j.ijhydene.2018.08.166.

[33] Zhao G, Fan J, Zhang H, Zhang Q, Yang J. Materials Science & Engineering A Exceptional mechanical properties of ultra- fi ne grain AZ31 alloy by the combined processing of ECAP , rolling and EPT. Mater Sci Eng A 2018;731:54-60. doi:10.1016/j.msea.2018.05.112.

[34] David E. Nanocrystalline magnesium and its properties of hydrogen sorption 2007;20:87-90.

[35] Skryabina N, Medvedeva N, Gabov A, Fruchart D, Nachev S, Rango P De. Impact of Severe Plastic Deformation on the stability of MgH 2. J Alloys Compd 2015;645:S14-7. doi:10.1016/j.jallcom.2015.03.128.

[36] Huang S, Chiu C, Chou T, Rabkin E. ScienceDirect Effect of equal channel angular pressing (ECAP) on hydrogen storage properties of commercial magnesium alloy AZ61 2018;3:0-9. doi:10.1016/j.ijhydene.2018.01.044.

Chapter 9

Alloying Elements of Magnesium Alloys: A Literature Review

Nouha Loukil

Abstract

Magnesium alloys are the lightest structural metal. The lightness is the main reason for the interest for Mg in various industrial and clinical applications, in which lightweight structures are in high demand. Recent research and developments on magnesium Mg alloys are reviewed. A particular attention is focused on binary and ternary Mg alloys consisting mainly of Al, Zn, Mn, Ca and rare earth (RE) elements. The effects of different alloying elements on the microstructure, the mechanical and the corrosion properties of Mg alloys are described. Alloying induces modifications of the microstructural characteristics leading to strengthening mechanisms, improving then the ductility and the mechanical properties of pure Mg.

Keywords: magnesium alloy, alloying element, strengthening, mechanical properties

1. Introduction

Magnesium alloys are the lightest structural metal view to the very low density of 1.74 g/cm^3 and designed as Green Structure Metal [1]. Taking into account its very negative potential (−2.34 V), Mg is a reducing agent and is able to combine with oxygen, sulfur and halogen compounds. This reducing power finds its interest in the production of sacrificial anodes that prevents corrosion. Meanwhile, this reducing agent constitutes a major barrier to the use of Mg as a structural material. In addition to this undesirable property, poor wear resistance of pure Mg hinders its use for different applications.

That's why, pure Mg is combined with other metal elements to improve their properties even at high temperatures, namely manganese, aluminum, zinc, silicon, copper, zirconium and rare-earth metals. Mg alloys, non-ferrous material, are characterized with low density, high ductility, strength and acceptable corrosion resistance.

The lightness is the main reason for the interest in the civil and military transport sector for Mg, in which lightweight structures are required. When compared with metallic structure namely, aluminum Al and iron Fe, the density of Mg is much lower than those of these metals [2]. In reverse, Mg exhibits similar specific mechanical properties, mainly excellent castability and machinability compared to a metal which is durable [3, 4]. When used as alloying element in metallic material, Mg enhances the mechanical properties of Aluminum and the malleability of the iron.

Compared to that of metallic structures, Mg alloys show higher weight/strength ratio. They possess an elastic modulus of 45 GPa and tensile strength of 160–365 MPa [5]. Based on the above reasons, Mg alloys have been widely used in the aerospace industry, mechanic manufacture and automotive industry. Indeed, the replacement

IntechOpen

in the three major components (body, power train and chassis) of a vehicle by Mg alloys lead to weight reduction of 20–70% [6].

Mg alloys provides an excellent property of damping vibration and heat dissipation property which is an important factor for different automobile and aerospace industries. As well known, the vibration is a kind of loss and affects the efficiency of the vehicle.

Special attention is paid to Mg-based materials for clinical applications (orthopedic applications, critical wounds ...) owing to its density that is very close to that of human bone (1.75 g/cm^3), higher specific strength and low elastic modulus. Furthermore, Mg is biocompatible as it is essential for several biological reactions and as a co-factor for enzymes.

Quite opposite to the conventionally used metallic materials such as stainless steel and Ti alloys that exhibit stress shielding and metal ion releases, Mg is biodegradable. That is to say, Mg entirely degrades in the human body preventing then the need for second surgical procedure to remove the implants material [7]. This has received a widespread attention from the scientific and medical community [8]. However, the biodegradability of implanted Mg alloys is hindered by an accelerated degradation rate in chloride-abundant environments like human body [9]. Generally, the period of bone remodeling is about 3 to 6 months. To be wary of this suggestion, the rate of degradation must be controlled by suitable surface modification in order to enhance the duration of effectiveness of the implantable material.

Researchers have been working on synthesis and characterization of Mg-based biomaterials with a variety of composition in order to control the degradation rate of Mg that leads to a loss of mechanical properties and contamination in the body. The alloying elements affect the characteristics and performance of Mg alloys.

This paper is a comprehensive review that compiles the recent literature on the important alloying elements and their impacts on the properties of Mg alloys.

2. Designation and types of Mg alloys

According to the addition elements, Mg alloys are designated in different ways. Alloys are designated by letters corresponding to their main addition elements followed respectively by the percentage of each element. The American Society for Testing and Materials ASTM developed a method to designate Mg alloys which are named by their main alloying elements. The first two letters indicate the alloying elements used in the greatest quantity. One or two letters are followed by numbers which represent the percentage by weight of the elements rounded to the nearest whole number. The ASTM code for alloying elements is as follows, aluminum is designated by the letter A, zinc by the letter Z, manganese by the letter M, silicon by the letter S, yttrium by the letter W, zirconium by the letter K, silver by the letter Q, thorium by the letter H. The most common families are: AZ (example AZ31), AM (example AM60), AS (example AS41), WE (example WE43) and AE (example AE42). For instance, AZ91 Mg alloy contains 9% of Al and 1% of Zn and the rest by pure Mg.

Based on the process of operation, Mg alloys can be categorized into two groups: cast alloys and wrought alloys.

2.1 Cast Mg alloys

Cast alloys are basically made by pouring the molten liquid metal into a mold, within which it solidifies into the required shape. Depending on the chemical composition, there are two groups of cast Mg alloys [10].

The first group includes Mg-Al alloys in which Al amount does not exceed 10% with an addition of Zn and Mn. These alloys are characterized by a low cost of manufacture. Indeed, this criterion is recommended for the industrialization and the commercialization of high performance of Mg alloys. The most commonly used is AM50 alloy mainly for die casting, and AZ91 for sand and die casting method. Meanwhile, their disadvantage is a low operating temperature-below 120 °C. The aging hardenability of these Mg-Al based alloys, such as AM60 alloy, is relatively poor. That's why, these alloys are prepared by high pressure die casting with relatively high cooling rate. The strength, the castability, the workability, the corrosion resistance and the weldability of these commercial AZ91 and AM60 alloys are not satisfying, but can be improved by introducing alloying elements. For instance, AZ91 alloy's strength is relatively high; however, its ductility is not so good due to the high Al content. Whereas, AM60 alloy has high ductility, but its strength is relatively low. Various alloying elements such as Ce, Nd, Y, Si, Ca, Ti, B, Sr., Sb, Bi, Pr have been used to enhance the operating temperature and the mechanical properties of modified AZ91 alloy [11]. Among these alloying elements Ce, Nd, Y, Bi and Sb are effective to improve the tensile properties of AZ91 alloy. For the modified AM60 alloys, various alloying elements, Ti, Nd, Sn and Ce elements are relatively effective to further enhance the mechanical properties of AM60 alloy. For instance, the tensile strength of 280 MPa and elongation of 11% could be obtained [12].

The second group of alloys is free from Al, but containing mostly Zn, rhenium RE and Y with an addition of Zr. Such alloys are recommended for high temperature until 250 °C, but the cost of their production increases due to the cost of alloy additions. The common commercial WE43 and WE54 alloys can be cast by sand casting process with a low cooling rate. Accordingly, the mechanical properties are further improved by solution and aging treatment. These alloys are largely used as sand castings. The strength of WE54 is basically attained via precipitation strengthening. Depending on the aging temperature and time, the precipitating sequence in WE alloys has been reported to involve the formation of phases β″, β′, and β. The equilibrium β phase is isomorphic to the Mg_5Gd phase and is identified as a $Mg_{14}Nd_2Y$ phase [13].

The mechanical properties of the cast alloys are determined on poured test bars according to standard ASTM procedures.

2.2 Wrought alloys

Wrought alloys are destined to mechanical working, such as forging, extrusion and rolling operations to shaping. Al, Mn and Zn are also the main alloying elements. Wrought alloys of Mg are sorted into heat treatable and non-heat treatable alloys. The use of wrought Mg alloys is limited less than 10%. This is due to the poor cold workability of the hexagonal close packed (HCP) crystal structure of Mg, generating low formability at room temperature. New Mg alloys have been developed to enhance their strength by modifying the existing alloy [14] or by grain refinement. It was also reported that advanced processing, such as hot extrusion, rolling, forging are able to refine the microstructure and improve the mechanical properties of Mg alloys. Many commercial wrought Mg alloys have been developed, such as AZ system, ZK system... Compared with wrought Mg alloys, casting Mg alloys have economical advantages due to their shorter processing cycle. Therefore, casting Mg alloys obtain more incremental use of almost 90% of total application products.

Based on literature data, new Mg alloys with high strength can be developed when modified the present commercial cast and wrought alloys [15] by strengthening mechanisms: alloying, grain refinement, precipitation and texture strengthening effect. To overcome the weakness of pure Mg, different elements have to be alloying with pure Mg to obtain Mg alloys with desired mechanical properties. Mg alloys shows an excellent specific strength and stiffness with dimensional stability due to its hexagonal crystal structure and to its atomic size (about 320 nm). Much progress has been achieved in strengthening of Mg alloys through solid solution strengthening using different alloying elements. The effects of such elements on the microstructure and mechanical properties are described.

2.2.1 Aluminum

Mg alloys including Al are called Mg-Al binary alloys. Al is the most widely used alloying element of Mg alloys as structural materials [16]. Indeed, Al improves the tensile strength, the ductility and the castability of Mg alloys at temperature not exceeding 120 °C [17]. If the amount of Al varies from 1 to 9%, the grains size of Mg-Al alloys decreases, increasing accordingly the micro-hardness [18]. Zheng et al. [18] demonstrated that the average grain size of Mg-Al alloy falls from 3097 μm to 111 μm if Al amount increases from 1% to 9%. **Figure 1** shows the influence of the addition of Al on the size of α-Mg dendrites for Mg-Al binary alloys. It can be seen that with 1% of Al to pure the structure of Mg, α-Mg dendrites are transformed from columnar to equiaxial (**Figure 1**). With higher Al amount, α-Mg dendrites become more developed and the dendrite arms become finer [18]. According to Mg-Al binary alloy phase diagram [19], α-Mg dendrite firstly precipitates during the solidification process, and then expels the redundant Al solute. In fact, this precipitation releases latent heat, causing an increase of the melt temperature ahead of the solidification interface. It decreases the degree of super-cooling, and suppresses the growth of α-Mg dendrites, leading to the grain refinement [18]. When Al exceeds 2%, an eutectic microstructure involving α-Mg and $Mg_{17}Al_{12}$ will be formed along the grain boundaries. The $Mg_{17}Al_{12}$ phase improves the corrosion potential and reduces the corrosion rate [20]. However, the $Mg_{17}Al_{12}$ phase has a relatively low melting temperature (437 °C), giving arise to the microstructure instability over 120 °C. This occurrence related to the grain boundary sliding explains the degradation of the mechanical properties of Mg-Al alloys at elevated temperatures [21]. Moreover, the micro-hardness of the α-Mg matrix increases from 36.3 HV to 50.1 HV. This

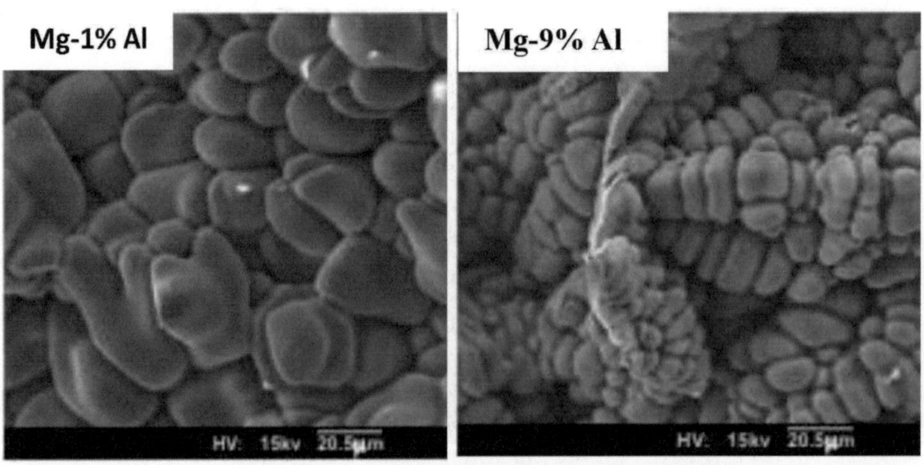

Figure 1.
Size and morphology of α-Mg dendrites of Mg-Al binary alloys with various Al amount [18].

increase is recognized to solution strengthening caused by Al solubilized in the α-Mg matrix.

The most popular alloys are Mg-Al-Zn (AZ) and Mg-Al-Mn (AM) as Zn improves the ambient temperature mechanical properties and Mn enhances the creep resistance of alloys [22].

2.2.2 Zinc Zn

Zinc is added in Mg alloys to enhance the tensile properties. Zinc decreases the weldability property [23]. Above 1% of Zn, it provides strengthening to Mg by solid solution [24]. If zinc is added in higher amount, the Mg alloy presents hot cracking and lower ductility [23]. In addition, micro porosity of sand casted Mg alloys is observed with an Al amount ranging from 2% to 10%. The dissolution of Zn in Mg diminishes its reducing power, improving then its oxidation resistance [25]. The corrosion rate of Mg-Zn alloys decreases with increasing Zn in Mg matrix. For instance, the corrosion rate of Mg-3%Zn alloy is 34.6% lower than that of pure Mg compact. Zhang et al. [26] reported that 6% of Zn diminishes the corrosion rate of the Mg alloy for implant applications.

The micro structural analysis indicates smaller grain size at higher Zn content. The reason for the grain sizes refinement is recognized to close-packed hexagonal structure of both Mg and Zn metals. The diffusion rate of Zn atoms in Mg matrix is fast, favoring an easier diffusion into Mg matrix and forming Mg solid solution or intermetallic compounds. **Figure 2** shows surface scanning micrographs of Mg-Zn alloys with various Zn content. The phases present in Mg-Zn alloys depend on the Zn amount. **Figure 2** shows some white phase along the Mg grain boundary. This white phase increases with higher Zn content, hindering both the movement of the grain boundary and the grain growth [27]. For instance, Mg-3%Zn alloy is mainly composed of α-Mg phase, whereas Mg-4%Zn alloy is composed of α-Mg and $MgZn_2$ phases [27]. The micro-hardness for different Mg-Zn alloys continuously increases with increasing Zn content. For illustration, the micro-hardness HV of Mg-3%Zn alloy is about 45% higher than those of pure Mg samples.

Mg-Zn based alloys are the most popular wrought Mg alloys with good room temperature strength and ductility. Recent researches attempt to develop new type wrought Mg-Zn alloys using the addition of alloying elements including RE (rare earth), non-toxic Ca, Sn and Mn to optimize the properties of Mg-Zn alloys at room and high temperatures. Mg-Zn-Zr (ZK) is the strongest system owing to good strength and elongation at room temperature.

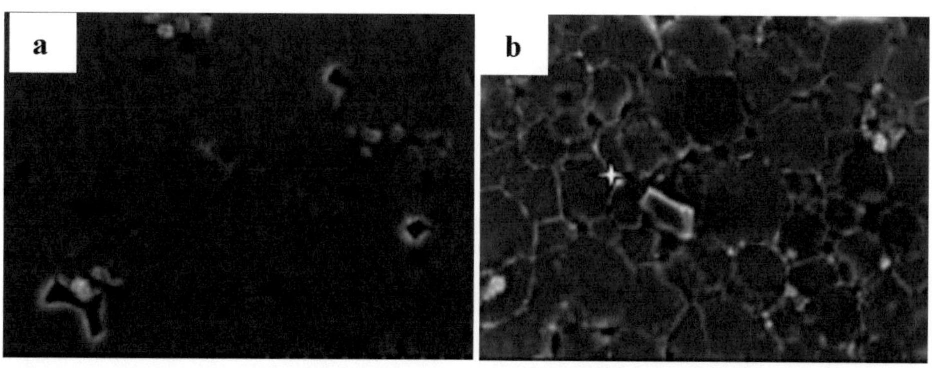

Figure 2.
SEM images of Mg-Zn alloys with different Zn contents: (a) 1% Zn and (b) 4% Zn [27].

The use of Mg-Zn alloys in medical applications as bio degradable materials is one of the research areas [28]. It is well known that the adding element Zn is one of the indispensable trace elements in the human body that promotes the growth, stimulates healing and participates in enzyme synthesis.

2.2.3 Manganese

The alloying element Mn improves the corrosion resistance of Mg alloy and limits the presence of harmful cathodic impurities such as Fe, Ni by the formations of intermetallic compound .... As matter of fact, such impurities lead to galvanizing oxidation of Mg [29]. Mn combines with impurities in order to moderate the corrosion of Mg-Al alloys. In the presence of Al and Fe, adding Mn produces an intermetallic phase $Al_8(Mn,Fe)_5$ that moderate the corrosion rate caused by Fe [30]. Until now, the exact Mn content addition required to counter-act the detrimental effect of the Fe impurity are still unknown.

With Al, the limit of solubility of Mn in the solid solution of Mg decreases, the strengthening by solid solution induced by Mn remains limited. That's why, the addition of Mn does not improve the mechanical properties of Mg. When investigated the effect of adding Mn into Mg-Gd alloys, Zhao et al. [31] demonstrated that the strength of alloys gradually increases while the ductility deteriorates. The main reasons are related to the combination of fine-grained strengthening, precipitation strengthening and texture strengthening [31]. In this regard, Cho et al. [32] stated that Mn refines grains of Mg-4Zn-0.5Ca alloy. This occurrence is due to the solute at the S/L interface aggregates in presence of Mn element, resulting different degrees of structural overcooling. Based on the biosafety to the human body, Mn can be accepted by the human body [32].

2.2.4 Ca

Incorporating Ca in Mg alloys can improve the mechanical properties and the corrosion behavior of Mg-Ca alloys. Ca is considered as an alloying element to develop Mg alloys for biomedical applications owing to their good biocompatibility.

As well known, Ca accelerates the bone growth. Moreover, cytocompatibility evaluation results indicated that Mg-1%Ca alloy induces no toxicity to human cells. In this regard, Li et al. [33] investigated the biodegradability within bone of Mg-Ca alloys with various Ca amounts ranging from from 1 to 20%. It was reported that 20% of Ca content makes Mg alloy very brittle.

Mg-Ca alloys with 0.6–1% Ca were reported to exhibit good mechanical properties and corrosion resistance [34]. The elongation of Mg-Ca alloy samples decreases with rising Ca content.

Previous studies suggested that Mg-Zn-Ca alloy system is a promising candidate for biodegradable implants in biomedical applications. With the increase of Ca content, the yield strength of Mg-Zn-Ca alloy increases. The addition of Ca to Mg-6%Zn alloy inhibits dynamic recrystallization and grain growth. Microstructural results indicate that Mg-Zn-Ca alloys consist of α-Mg matrix and $Ca_2Mg_6Zn_3/Mg_2Ca$ intermetallic phase mainly distributed along grain boundary [35].

High Ca content improves mechanical properties, while it is detrimental to corrosion resistance due to the micro-galvanic corrosion acceleration [36]. Literature data revealed that Mg-5%Zn-1%Ca exhibits excellent corrosion resistance and good biocompatibility.

2.2.5 Yttrium Y

Generally, the addition of Yttrium (Y) aims to improve elevated temperature plasticity and the creep resistance of Mg alloys. Wu et al. [37] studied the mechanical properties of pure Mg and binary Mg-Y alloys. They demonstrated that the elongation is proportional with the Y content, but the strength decreases. Adding Yttrium to Mg-Zn alloys is effective to weaken and change the basal texture of wrought Mg alloys [38].

When added to the cast alloy ZK60, the mechanical properties and ductility are enhanced at elevated temperature. This is ascribed to the formation of ternary Mg-Zn-Y phase with high thermal stability [39]. Meanwhile, alloys containing yttrium are expensive.

2.2.6 Rare earths RE elements

The rare Earths elements are common incorporating elements in Mg alloys. A large number of rare-earth elements have been investigated and proven to successfully refine the crystals and to enhance the creep and the corrosion resistance at elevated temperatures required for automobile engineering. Due to their high solubility in Mg, RE elements strengthen Mg alloys either by solid solution strengthening or precipitation hardening mechanisms [36]. For instance, the addition of RE elements to Mg-Al alloys accelerate the formation of the thermally stable (Mg,Al)xREy phases to improve the high temperature mechanical properties of wrought Mg alloys.

Depending on the chemical composition, various amounts of precipitates can be observed. **Figure 3** exhibits small precipitates containing Mg, Zn and O in Mg-1.5Zn (**Figure 3a**), while precipitates in the RE-containing alloys consist of Mg, Zn and RE elements [27]. The amount of precipitates of Mg-1.5Zn-0.2Gd alloys are more than Mg-1.5Zn alloy, leading to smaller grain size of this alloy. Compared with other rare earth elements, Gd shows higher solubility in Mg, allowing simultaneous solid solution hardening and precipitation strengthening, enhancing then the thermal stability of the microstructure in Mg alloys [27]. Yang et al. [39] pointed out that Gd acts as grain refinement and grain boundary strengthening. In this regard, Liu et al. [40] reported that Mg-1.5Zn-0.2Gd alloy exhibits higher plasticity with the elongation of 27%. Meanwhile, Gd simultaneously reduces the toughness [36]. These Mg alloy systems containing large amounts of rare-earth for solid solution strengthening result high costs for many practical applications.

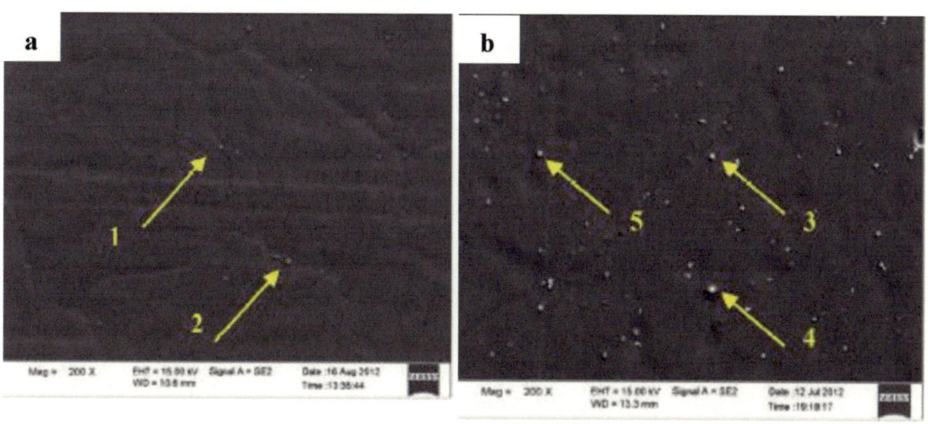

Figure 3.
Microstructural (SEM) of Precipitates in Mg alloy (a) Mg-1.5%Zn; (b) Mg-1.5%Zn-0.2%Gd.[27].

3. Conclusion

The attractive properties of Mg alloys include light weight, high specific strength, excellent castability and machinability .... However, these alloys have limited formability, limiting their industrial application. Strengthening of Mg alloys via the introduction of solid solution atoms and grain refinement additives are effective approaches. Much progress has been achieved in the development of high strength Mg alloys through solid solution strengthening using various types of solute atoms. Researchers incorporated various alloying elements such as Al, Zn, Ca, Mn and RE elements in Mg matrix. The chemical composition of Mg alloys affects the microstructure and then improves the mechanical properties and corrosion resistance.

The most commercial are the casting magnesium alloys such as AZ91, AM60 and WE43 due to their low cost. Meanwhile, the Mg alloy systems containing rare-earth and noble metal elements for solid solution strengthening result high costs. The heat treatment and the grain refinement via severe plastic deformation methods represent new approach to further strengthening of Mg alloys.

Conflict of interest

The authors declare no conflict of interest.

Author details

Nouha Loukil
Laboratory of Materials Engineering and Environment, Sfax, Tunisia

*Address all correspondence to: nloukil87@gmail.com

References

[1] Yang M, Liu Y H, Jin H Z, Song Y L. Influence of solidsolution treatment on tensile properties of cast AM50 magnesium alloy after corrosion tests. Transactions of Materials and Heat Treatment (in Chin.), 2012, 33(7): 45

[2] G. Neite, K. Kubota, K. Higashi et F. Hehmann, Livre "Structure and Properties of Nonferrous Alloys", VCH, 1996

[3] Norme "Standard Test Methods for Tension Testing of Metallic Materials",E8, A. S. T. M. International, 2004

[4] A. S. M. I. H. Commitee, General Introduction, Livre "ASM Speciality Handbook: Aluminum and Aluminum Alloys", A.S.M. International, 1993, pp. 3-17

[5] Critical assessment of strengthening mechanism of magnesium alloys:Review Brijesh Prasad, Bhingole P P

[6] Magnesium and Its Alloys in Automotive Applications – A Review, D. Sameer Kumar1, C. Tara Sasanka1, K. Ravindra1 , K.N.S. Suman

[7] Poly caprolactone/titanium dioxide nanofiber coating on AM50 alloy for biomedical application

[8] Jin, W., Wang, G., Qasim, A. M., Mo, S., Ruan, Q., Zhou, H., et al. (2019). Corrosion protection and enhanced biocompatibility of biomedical Mg-Y-RE alloy coated with tin dioxide. Surf. Coat. Technol

[9] J. Gray-Munro, M. Strong , J. Biomed. Mater. Res. A 90 (2009) 339-350

[10] Friedrich H, Mordike B. Magnesium Technology. Berlin Heidelberg:Springer-Verlag; 2006. 677 p

[11] Yang, Y., Liu, Y., Qin, S., & Fang, Y. (2006). High cycle fatigue properties of die-cast magnesium alloy AZ91D with addition of different concentrations of cerium. Journal of Rare Earths, 24(5), 591-595.

[12] F. He, J.P. Li, Z. Yang, Special Casting and Nonferrous Alloys 28 (2008) 227. 229 (in Chinese)

[13] Kiełbus, A. (2018). Microstructure and properties of casting magnesium alloys designed to work in elevated temperature. In Magnesium Alloys-Selected Issue. IntechOpen.

[14] J.F. Nie, Y.M. Zhu, A.J. Morton, Metall. Mater. Trans. A 45A (2014) 3338-3348

[15] T.Y. Kwak, H.K. Lim, W.J. Kim, J. Alloys Compd. 658 (2016) 157-169

[16] Recent research and developments on wrought magnesium alloys, Sihang You, Yuanding Huang, Karl Ulrich Kainer, Norbert Hor

[17] M.S. Dargusch, K. Pettersen, K. Nogita, M.D. Nave, G.L. Dunlop, Mater. Trans. 47 (2006) 977-982

[18] Zheng Wei-chao., LI Shuang-shou., TANG Bin., & ZENG Da-ben. (2006) Microstructure and properties of Mg-Al binary alloys. China Foundry, Vol.3, No.4, 270-274

[19] YU Jue-qi, YI Wen-zhi, CHEN Bang-di, et al. Phase Diagrams of Binary Alloys. Scientific & Technical Press, Shanghai, 1987. (in Chinese)

[20] Young J K, Chang D Y, Jong D L, et al. Effect of Mg17Al12 precipitate on corrosion behavior of AZ91D magnesium alloy. Materials Science Forum, 2003, 419-422: 851-819

[21] F. Kabirian, R. Mahmudi, Metall. Mater. Trans. A 41A (2010) 3488-3498

[22] A.A. Luo, A.K. Sachdev, Metall. Mater. Trans. A 38A (2007) 1184-1192

[23] Avedesian.; Michael M.; Hugh Baker.; eds. ASM specialty handbook:magnesium and magnesium alloys. ASM international, 1999

[24] C. Shaw et H. Jones, The contributions of different alloying additions to hardening in rapidly solidified magnesium alloys, Materials Science and Engineering: A, 226-228, 1997, pp. 856-860., 59

[25] D. Daloz, P. Steinmetz et G. Michot, Corrosion behavior of rapidly solidified magnesium-aluminum-zinc alloys, Corrosion, 53, 1997, pp. 944-954

[26] Zhang, B., Hou, Y. L., Wang, X. D., Wang, Y., and Geng, L. (2011). Mechanical properties, degradation performance and cytotoxicity of Mg–Zn–Ca biomedical alloys with different compositions. Mater. Sci. Eng. C 3, 1667-1673.

[27] Jingyuan, Y., Jianzhong, W., Qiang, L., Jian, S., Jianming, C., & Xudong, S. (2016). Effect of Zn on microstructures and properties of Mg-Zn alloys prepared by powder metallurgy method. Rare Metal Materials and Engineering, 45(11), 2757-2762

[28] Shaoxiang Zhang. et al.., (2010). Research on an Mg–Zn alloy as a degradable biomaterial. Acta Biomaterialia , Volume 6, Issue 2, Pp. 626-640.

[29] M. Colombié, Chapitre 13 - Magnésium et alliages de magnésium, Livre "Matériaux Métalliques", Dunod, 2000, pp. 679-726.

[30] M. Perguleryuz, Chapter 5 - Alloying behavior of magnesium and alloy design, Livre "Fundamentals of magnesium alloy metallurgy", Woodhead Publishing, 2013, pp. 152-196.

[31] Effect of manganese on microstructure and properties of Mg-2Gd magnesium alloy, Tianshuo Zhao, Yaobo Hu, BingHe, ChaoZhang, Tianxu Zheng, FushengPanab

[32] Cho, D. H., Lee, B. W., Park, J. Y., Cho, K. M., and Park, I. M. (2017). Effect of Mn addition on corrosion properties of biodegradable Mg-4Zn-0.5Ca-xMn alloys. J. Alloys. Compd. 695, 1166-1174

[33] Yin, P., Li, N. F., Lei, T., Liu, L., & Ouyang, C. (2013). Effects of Ca on microstructure, mechanical and corrosion properties and biocompatibility of Mg–Zn–Ca alloys. Journal of Materials Science: Materials in Medicine, 24(6), 1365-1373

[34] Wan Y, Xiong G, Luo H, He F, Huang Y, Zhou X. Preparation and characterization of a new biomedical magnesium–calcium alloy. Mater Des. 2008;29 (10):2034-7

[35] Han, Y. Y., You, C., Zhao, Y., Chen, M. F., & Wang, L. (2019). Effect of Mn Element Addition on the Microstructure, Mechanical Properties, and Corrosion Properties of Mg-3Zn-0.2 Ca Alloy. Front. Mater. 6: 324.

[36] J. Geng, X. Gao, X.Y. Fang, J.F. Nie, Scr. Mater. 64 (2011) 506-509; Two main and a new type rare earth elements in Mg alloys: A Review, Linghang Kong

[37] B.L. Wu, Y.H. Zhao, X.H. Du, Y.D. Zhang, F. Wagner, C. Esling, Mater. Sci. Eng. A 527 (2010) 4334-4340

[38] D. Wu, R.S. Chen, W.N. Tang, E.H. Han, Mater. Des. 41 (2012) 306-313

[39] Shi LL, Huang Y, Yang L et al (2015) Mechanical properties and corrosion behavior of Mg-Gd-Ca-Zr alloys for

medical applications, J Mech Behav Biomed 47:38-48

[40] The effect of Y, Ce and Gd on texture, recrystallization and mechanicalproperty of Mg–Zn alloys Peng Liu, Haitao Jiang, Zhengxu Cai, Qiang Kang, Yun Zhang